FEU.

DEGRADATION OF LIGNOCELLULOSICS
IN RUMINANTS AND IN INDUSTRIAL PROCESSES

Proceedings of a workshop held in Lelystad (The Netherlands) from 17 to 20 March 1986 under the auspices of COST (European Cooperation in Scientific and Technical Research)—COST 84-bis, organised in collaboration with the Commission of the European Communities by

—the Institute for Livestock Feeding and Nutrition Research (IVVO), Lelystad

and

—the Institute for Storage and Processing of Agricultural Products (IBVL), Wageningen

DEGRADATION OF LIGNOCELLULOSICS IN RUMINANTS AND IN INDUSTRIAL PROCESSES

Edited by

J. M. VAN DER MEER

Institute for Livestock Feeding and Nutrition Research (IVVO),
Lelystad, The Netherlands

B. A. RIJKENS

Institute for Storage and Processing of Agricultural Products (IBVL),
Wageningen, The Netherlands

and

M. P. FERRANTI

Commission of the European Communities, Brussels, Belgium

ELSEVIER APPLIED SCIENCE
LONDON and NEW YORK

ELSEVIER APPLIED SCIENCE PUBLISHERS LTD
Crown House, Linton Road, Barking, Essex IG11 8JU, England

Sole Distributor in the USA and Canada
ELSEVIER SCIENCE PUBLISHING CO., INC.
52 Vanderbilt Avenue, New York, NY 10017, USA

WITH 24 TABLES AND 18 ILLUSTRATIONS

British Library Cataloguing in Publication Data

Degradation of lignocellulosics in ruminants
and in industrial processes.
1. Waste products as feed
I. Meer, J. M. van der   II. Rijkens, B. A.
III. Ferranti, M. P.
636.08′552   SF99.W34

ISBN 1-85166-165-4

Library of Congress CIP data applied for

Publication arrangements by Commission of the European Communities, Directorate-General Telecommunications, Information Industries and Innovation, Luxembourg

EUR 11084

Printed in Great Britain by Galliard (Printers) Ltd, Great Yarmouth

Contents

INTRODUCTION

The workshop reported in this volume was sponsored by the Commission of the European Communities, Directorate-General for Science, Research and Development (DG XII), under the Concerted Action Project COST 84-bis, entitled "Use of lignocellulose-containing by-products and other plant residues for animal feeding". The papers were circulated among the participants in advance, and two discussion leaders, Professor A.J.H. van Es and Dr. F. Rexen each introduced the three sessions.

Within the European Community and the adjacent countries there is a structural overproduction of certain primary and secondary agricultural products (such as soft wheat, beet sugar, various kinds of straw and other by-products and dairy products) which in several cases involve the Community in high costs. On the other hand, there are deficiencies of other products, such as papermaking fibre, feed protein and certain kinds of agro-based chemicals and fuel. A shift to the production of "industrial" and "energy" crops is therefore to be recommended.

Both the primary production and the processing of agricultural products lead to various by-products which are often used as feed. The lignocellulosic by-products, however, have a low feeding value. They often occur in surplus in particular areas, and so become wastes. A shift towards industrial and energy crops will lead to further lignocellulosic by-products for which uses will have to be found in order to prevent them becoming wastes.

If the two major components, lignin and cellulose, can be separated, the intrinsic value of these low grade lignocellulosics as an animal feed, or as raw materials for the production of protein, chemicals and fuels, can be raised. The microbial breakdown which occurs reasonably effectively in the rumen may be a good basis for the industrial utilization of lignocellulosics. Increasing knowledge of the influence of the structure of the cell walls and the effects of chemical and mechanical pretreatments may open new avenues for future industrial processing.

These new insights mainly concern straw, but they may also help develop new industrial markets for the various other wasted or under-utilized lignocellulosics, such as cereal, rape and seed (grass) straw, sunflower and corn stalks, reeds, wood wastes, bark, press cakes or residues of olives, grapes , bagasses etc.

In feed technology, relatively simple and cheap pre-treatments (using alkali, SO_2, steam, mechanical treatment, and fungi) are being tried in order to improve the digestibility or the feeding value of the whole product (straw for example) so that it can be fed effectively to ruminants. These kind of pre-treatments can generally be applied on a small scale at farm level.

More complete separation of the lignin and the cellulose components requires more complicated technological procedures which can produce final products of higher added value and for a broad spectrum of applications. Such processes are under development by the industry.

THE CHEMISTRY OF LIGNOCELLULOSIC MATERIALS
FROM AGRICULTURAL WASTES IN RELATION TO PROCESSES
FOR INCREASING THEIR BIODEGRADABILITY

R. D. Hartley

Animal & Grassland Research Institute,

Hurley, Maidenhead, Berks. SL6 5LR, UK

ABSTRACT

This review is concerned mainly with graminaceous wastes, particularly cereal straws. The chemical constitution of cell walls before and after processing to increase biodegradability is discussed. New methods of processing are referred to.

INTRODUCTION

Much recent work on agricultural wastes has concentrated on graminaceous materials, due mainly to the large surpluses of cereal straw in several EEC countries, in the UK, for example, the amount of straw produced is *ca.* 15 million tonnes per annum. In addition to wheat, barley, oat, rice and rye straws, poor quality grass hays and residues from the production of maize grain are receiving attention.

It is intended in this review to concentrate on recent developments particularly at this Institute.

CHEMICAL CONSTITUTION OF PLANT CELL WALLS

In graminaceous wastes, cell walls account for a large proportion of the dry matter. For example, in wheat, barley and oat straws the proportion often exceeds 80% (Table I; Hartley, Mason, Keene and Cook, to be published). The walls are mainly composed of cellulose, hemicellulose and aromatic materials (including lignin) together with minerals (including silica) and small amounts of nitrogen-containing materials (Table I). In our studies of wheat and barley straws a nitrogen content of *ca.* 0.2% in the cell wall fraction is common (see Jackson, 1977, for a review of the cell wall constituents of other lignocellulosic materials, including legume and sugarcane wastes and woods). Recent reviews (Theander, 1985; Wilkie, 1985) discuss the chemical structure of the polysaccharide which is mainly comprised of cellulose and xylans, the latter having low proportions of L-arabinofuranosyl

Table I. The chemical composition of cereal straws and hays before and
after treatment with ammonia

Cereal straw or hay	In vitro organic matter digestibility (% dry matter)	Cell walls (% dry matter)	Ash (% dry matter)	Biodegradability of cell walls (% cell walls)
Wheat straw (var. Avalon)	43.4	86.5	5.9	22.7
Wheat straw after NH_3 treatment	57.6	79.2	6.1	31.3
Oat straw (var. Pennal)	48.7	84.8	6.3	22.2
Oat straw after NH_3 treatment	63.9	77.5	6.0	39.1
Ryegrass hay	58.2	66.2	6.4	26.6
Ryegrass hay after NH_3 treatment	68.0	59.9	6.7	50.9
Cocksfoot hay	64.5	62.3	7.7	32.6
Cocksfoot hay after NH_3 treatment	72.1	60.9	7.7	58.3

Cereal straw or hay	α-Cellulose (% cell walls)	Hemicellulose (% cell walls)	Lignin (% cell walls)	Ferulic plus p-coumaric acids (mg g^{-1} cell walls)
Wheat straw (var. Avalon)	45.9	45.0	10.5	7.7
Wheat straw after NH_3 treatment	48.6	38.9	11.6	4.0
Oat straw (var. Pennal)	48.0	42.0	9.9	11.1
Oat straw after NH_3 treatment	51.5	34.7	10.8	5.6
Ryegrass hay	43.7	35.4	8.0	9.0
Ryegrass hay after NH_3 treatment	47.6	27.3	9.4	3.2
Cocksfoot hay	47.1	48.0	11.4	8.7
Cocksfoot hay after NH_3 treatment	46.0	47.3	15.5	2.8

In vitro organic matter digestibility was determined by the rumen liquor-pepsin method (Tilley and Terry, 1963). Cell walls were determined by a modification (Hartley, Jones and Fenlon, 1974) of the method of Van Soest and Wine (1967). Biodegradability of cell walls was determined with a commercial 'cellulase' and lignin by a modified Klason procedure (Hartley and Jones, 1978). Ferulic and *p*-coumaric acids (*cis* plus *trans* isomers) were determined by a modification (Hartley, in press) of the method of Hartley and Buchan (1979). α-Cellulose and hemicellulose were determined by a modification (Hartley, Mason and Keene, to be published) of the method of Thornber and Northcote (1961).

---

units linked to 0-3 of the D-xylopyranosyl units. The hemicellulose fraction including xylans also contains 1 to 2% acetyl groups (Bacon and Gordon, 1980; Theander, Udén and Åman, 1981). The reviews of Theander and Wilkie also discuss the variations in chemical constitution that occur with species, variety and time of harvest (see also Theander and Åman, 1978, 1984; Åman and Nordkvist, 1983). From such work, it is clear that plant breeding techniques could greatly increase the yield of straw biomass if an increased demand arose. Plant breeding might also lead to the production of straws with increased biodegradability.

Estimation of the proportion of lignin (polymeric materials derived from phenyl propane units) in cell walls by the Klason or other techniques is very crude, as discussed earlier (Hartley, 1981). Nevertheless, the amounts of phenolic acids (mainly *p*-coumaric and ferulic acids) associated with the cell walls and released by treatment with alkali can be determined with accuracy by high performance liquid chromatography (Hartley and Buchan, 1979; Hartley, in press). They constitute about 1 to 3% of graminaceous cell walls.

It has recently been established that, in cell walls of barley straw, the *trans* isomers of both *p*-coumaric and ferulic acids are ester-linked to arabinoxylan:enzymic degradation of the walls released the compounds shown in Figure 1 (Mueller-Harvey, Hartley, Harris and Curzon, in press). Kato, Azuma and Koshijima (1983) had earlier shown that FAXX (Figure 1) could be obtained from bagasse by enzymic degradation. Mueller-Harvey *et al.* calculated that 1 in every 31 arabinose units in the walls was esterified with *p*-coumaric acid and 1 in every 15 with ferulic acid. The amounts of PAXX (Figure 1) and FAXX released accounted for one sixth of *p*-coumaric acid and half of the ferulic acid released from the walls by treatment with sodium hydroxide. It seems likely that if the walls had been ground more finely, then more PAXX and FAXX would be released. Earlier work (Hartley and Jones, 1978; Hartley and Haverkamp, 1984) had suggested that *p*-coumaric acid was

6

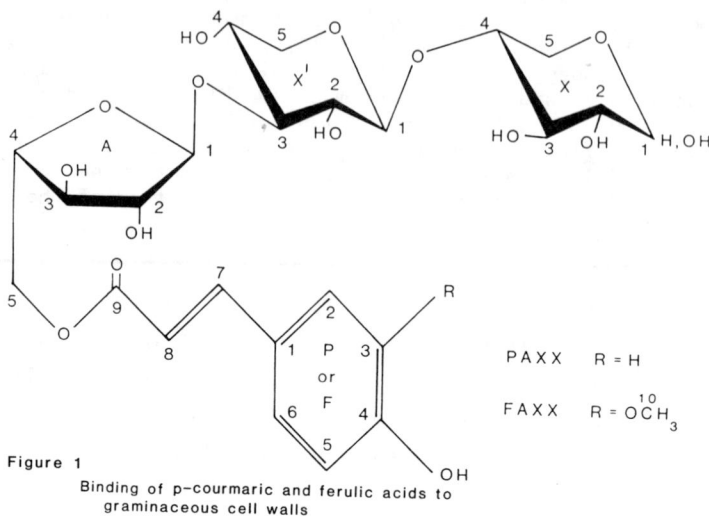

Figure 1

Binding of p-courmaric and ferulic acids to
graminaceous cell walls

PAXX   R = H

FAXX   R = OCH₃

more difficult than ferulic acid to remove from cell walls. Preliminary
studies with other graminaceous species suggest that they contain similar
linkages of *p*-coumaric and ferulic acids to arabinoxylans (Hartley and Keene,
to be published).

It has been proposed (Markwalder and Neukom, 1976; Hartley and Jones,
1976) that some ferulic acid units attached to xylan chains react by oxida-
tive coupling to form diferulic acid which remains associated with the hemicellulose fraction. This acid which has been identified in grasses and cereals
has a molecular weight twice that of ferulic acid minus two (removal of two
hydrogen atoms). It is possible that the ferulic and *p*-coumaric acid units
could be precursors of more complex aromatics ('lignin') associated with cell
walls.

Silica is an important constituent of the cell walls of some tropical
graminaceous crops, for example accounting for up to 15% of rice straw. This
is in contrast to the much smaller amounts present in wheat, barley, oats and
rye.

An earlier review (Hartley, 1985) discussed the application of nuclear
magnetic resonance, near infrared reflectance and pyrolysis-mass spectrometry
to the chemical characterisation of plant cell walls. These techniques are
likely to become of increasing importance as a means of unravelling the
complex structures of cell walls (Himmelsbach and Barton, 1980; Barton and
Burdick, 1983; Hartley and Haverkamp, 1984).

TREATMENT OF LIGNOCELLULOSIC GRAMINACEOUS WASTES TO INCREASE THEIR BIODEGRADABILITY

Recent reviews and books (including Sundstøl, 1981, 1984; Sundstøl and Owen, 1984; Hartley, 1981, 1985) have discussed various chemical and physical processes for increasing the biodegradability of cereal straws and other lignocellulosic wastes. The most widely used processes involve alkali treatment which is commonly used on farms to upgrade straws. Two important systems involve the application of anhydrous ammonia to straw (3.5 parts of NH_3 to 100 parts dry straw) and either heating at $ca.$ 90° in an oven ('the oven method') or leaving in stacks for $ca.$ six weeks at ambient temperature ('the stack method'). In the UK, the 'stack method' is often based on the use of aqueous solutions of ammonia to avoid transportation of the anhydrous form of the chemical. The increase in digestibility resulting from treating straws or hays with aqueous ammonia has sometimes been unexpectedly low. This may be explained by microbial production of carbon dioxide in closed stacks followed by reaction of the gas with the added ammonia to form ammonium carbamate/carbonate (Mason, Wilson, Keene and Hartley, 1985). Danish workers have recently patented the use of ammonium bicarbonate instead of ammonia for use in the oven process. This Institute has recently applied for a patent for the use of solid ammonium salts plus alkali (e.g. ammonium sulphate plus quicklime) to produce NH_3 in $situ$ as a safe, economical means of upgrading cereal straws and poor quality hays on the farm (Mason et $al.$, in press). Our philosophy is that techniques for upgrading lignocellulosic materials must be both economical and safe. For the production of animal feed, we prefer treatments that can be undertaken on farms as the transport of wastes to an industrial plant adds considerably to processing costs.

A major effect of alkali treatment is to increase considerably the biodegradability of the cell walls (see Table I). Recently it has been shown (Hartley, Mason, Keene and Cook, to be published) that the relationship between the amount of phenolics (including p-coumaric and ferulic acids) released by alkali treatment and the biodegradability of the cell walls is linear for a particular variety of cereal straw or grass hay. However, variation between species is large, suggesting that other factors are also important in decreasing biodegradability.

Sodium hydroxide is more effective than ammonia in increasing the biodegradability of graminaceous straws but it has proved difficult to devise suitable methods for use directly by the farmer. The addition of sodium to animal diets has also caused some concern while the addition of nitrogen (from ammonia treatment) to straw-based diets can be beneficial.

Other chemicals being investigated for straw treatment for animal feed include sulphur dioxide, ozone and alkaline hydrogen peroxide (Ben-Ghedalia and Miron, 1981; Ben-Ghedalia, 1984; Kerley *et al.*, 1985). Such techniques could lead to new industrial uses of lignocellulosic wastes. Some of these processes cleave benzene rings and hence should more effectively 'delignify' lignocellulosic materials.

A new process (Deschard, Mason and Tetlow, 1984) for the production of animal feed involves the harvesting of whole crop cereal without separation into straw and grain, followed by ensilage with alkali (e.g. sodium hydroxide). The crop is preserved by this process and, at the same time, the straw component is upgraded. Trials with beef cattle have given large increases in live-weight.

There is considerable interest, in the EEC, in possible industrial uses for whole crop cereals (grain plus straw) as a means of utilising surpluses (Rexen and Munck, 1984; Rexen, 1985). One suggestion is to set up integrated agricultural-industrial units capable of producing several different products for both industrial use (e.g. starch, protein) and animal feeds (e.g. high protein residues from industrial processes). It is acknowledged that technical and political problems need to be solved before such units could become economically viable.

In the future, cheaply produced enzymes (lignases, polysaccharidases) could play an important role in the treatment of lignocellulosic material in industrial or on-farm processes. They could eventually displace chemical treatments as they are less likely to lead to the production of undesirable by-products.

POSSIBLE HAZARDS ASSOCIATED WITH THE TREATMENT OF LIGNOCELLULOSIC WASTES

It is necessary to ensure the health and safety of farm workers when treating lignocellulosic materials on the farm for the production of animal feed, and to ensure that the processed materials are not detrimental to animal or human health. One factor that requires investigation is the possible formation, when lignocellulosic materials containing soluble sugars are treated with ammonia at elevated temperatures, of 4-methyl-imidazole, a compound which is toxic to animals (Nishie, Waiss and Keyl, 1969).

ACKNOWLEDGEMENTS

The author wishes to thank R. J. Barnes for determinations of *in vitro* organic matter digestibility. The Animal and Grassland Research Institute is funded through the Agricultural and Food Research Council.

REFERENCES

Bacon, J. S. D. and Gordon, A. H. 1980. Effects of various deacetylation procedures on the nylon bag digestibility of barley straw and of grass cell walls recovered from sheep faeces. J. Agric. Sci. Camb., 94, 361-367.

Barton, F. E. and Burdick, D. 1983. Prediction of forage quality with NIR reflectance spectroscopy. Proc. XIV Int. Grassld Congr. (Smith, J. A. and Hays, V. W. eds) pp. 532-534. Westview Press, Boulder, USA.

Deschard, G., Mason, V. C. and Tetlow, R. M. 1984. Preliminary studies of intake and growth in steers fed whole crop wheat treated with various chemicals. Seventh Silage Conference, Queen's University, Belfast, 4-6 September, pp. 55-56.

Hartley, R. D. 1981. Chemical constitution, properties and processing of lignocellulosic wastes in relation to nutritional quality for animals. Agric. Environm., 6, 91-113.

Hartley, R. D. 1985. Chemistry of lignocellulosic plant materials and non-microbial processes for increasing their feed value for the ruminant. Improved utilisation of lignocellulosic materials in animal feed. OECD, Paris, pp. 10-30.

Hartley, R. D. In press. HPLC for the separation and determination of phenolic compounds in plant cell walls. In Linskens, H-F and Jackson, J. F. (ed.) Modern Methods of Plant Analysis, Volume 4. Elsevier, Amsterdam.

Hartley, R. D. and Buchan, H. 1979. High-performance liquid chromatography of phenolic acids and aldehydes derived from plants from the decomposition of organic matter in soil. J. Chromatogr., 180, 139-143.

Hartley, R. D. and Haverkamp, J. 1984. Pyrolysis-mass spectrometry of the phenolic constituents of plant cell walls. J. Sci. Food Agric., 35, 14-20.

Hartley, R. D. and Jones, E. C. 1976. Diferulic acid as a component of cell walls of *Lolium multiflorum*. Phytochemistry, 15, 1157-1160.

Hartley, R. D. and Jones, E. C. 1978. Effect of aqueous ammonia and other alkalis on the *in vitro* digestibility of barley straw. J. Sci. Food Agric., 29, 92-98.

Hartley, R. D., Jones, E. C. and Fenlon, J. C. 1974. Prediction of the digestibility of forages by treatment of their cell walls with cellulolytic enzymes. J. Sci. Food Agric., 25, 947-954.

Himmelsbach, D. S. and Barton, F. E. 1980. C-13 nuclear magnetic resonance of grass lignins. J. Agric. Food Chem., 28, 1203-1208.

Jackson, M. G. 1977. Review article; the alkali treatment of straws. Anim. Feed Sci. Technol., 2, 105-130.

Kato, A., Azuma, J. and Koshijima, T. 1983. A new feruloylated trisaccharide from bagasse. Chem. Letters (Japan), 137-140.

Kerley, M. S., Fahey, G. C., Berger, L. L., Gould, J. M. and Baker, F. L. 1985. Alkaline hydrogen peroxide treatment unlocks energy in agricultural by-products. Science, 230, 820-822.

Markwalder, H. U. and Neukom, H. 1976. Diferulic acid as a possible cross-link in hemicelluloses from wheat endosperm. Phytochemistry, 15, 836-837.

Mason, V. C., Cook, J. E., Smith, T., Siviter, J. W., Keene, A. S. and Hartley, R. D. In press. The new AGRI-AM process for ammoniation. Stack ammoniation of mature forages using ammonium salts. ARNAB Workshop, Alexandria, Egypt.

Mason, V. C., Wilson, R. F., Keene, A. S. and Hartley, R. D. 1985. The effect of carbon dioxide on the efficiency of upgrading hay by ammoniation. Br. Soc. Anim. Prod. 40, 533-534.

Mueller-Harvey, I., Hartley, R. D., Harris, P. J. and Curzon, E. H. In press. Linkage of p-coumaroyl and feruloyl groups to cell wall polysaccharides of barley straw. Carb. Res., 147.

Nishie, K., Waiss, A. C. and Keyl, A. C. 1969. Toxicity of methylimidazoles. Toxicol. Appl. Pharm., 14, 301-307.

Rexen, F. 1985. Total industrial utilisation of the cereal plant. In Hill, R. D. and Munck, L. (ed.) Progress in Biotechnoloty, Volume 1, New Approaches to Research on Cereal Carbohydrates. Elsevier, Amsterdam, pp. 333-337.

Rexen, F. and Munck, L. 1984. Cereal crops for industrial use in Europe. Report prepared for the Commission of the European Communities. EUR 961EN.

Sundstøl, F. 1981. Methods for treatment of low quality roughages. In Kategile, E., Said, A. N. and Sundstøl (eds) Utilization of Low Quality Roughages in Africa. Lamport-Gilbert, Reading, UK, pp. 61-89.

Sundstøl, F. 1984. Ammonia treatment of straw: methods for treatment and feeding experience in Norway. Anim. Feed Sci. Technol., 10, 173-187.

Theander, O. 1985. Review of straw carbohydrate research. In Hill, R. D. and Munck, L. (ed.) Progress in Biotechnology, Volume 1, New Approaches to Research on Cereal Carbohydrates. Elsevier, Amsterdam, pp. 217-230.

Theander, O. and Åman, P. 1978. Chemical composition of some Swedish cereal straws. Swedish J. Agric. Res., 8, 189-194.

Theander, O. and Åman, P. 1984. Anatomical and chemical characteristics. In Straw and Other Fibrous By-products as Feed (Sundstøl, F. and Owen, E. eds) pp. 45-78. Developments in animal and veterinary sciences, 14. Elsevier, Amsterdam.

Theander, O., Udén, P. and Åman, P. 1981. Acetyl and phenolic acid substituents in timothy of different maturity and after digestion with rumen micro-organism or a commercial celluloase. Agric. Environm., 6, 127-133.

Thornber, J. P. and Northcote, D. H. 1961. Changes in the chemical composition of a cambial cell during its differentiation into xylem phloem tissue in trees. Biochem. J., 81, 449-455.

Tilley, J. M. A. and Terry, R. A. 1963. A two stage technique for the *in vitro* digestion of forage crops. J. Br. Grassl. Soc., 18, 104-111.

Van Soest, P. J. and Wine, R. H. 1967. Use of detergents in the analysis of fibrous feeds. IV. Determination of plant cell wall constituents. J. Assoc. Off. Anal. Chem., 50, 50-55.

Wilkie, K. C. B. 1985. The non-cellulosic polysaccharides of non-endospermic parts of grasses. In Hill, R. D. and Munck, L. (ed.) Progress in Biotechnology, Volume 1, New Approaches to Research on Cereal Carbohydrates. Elsevier, Amsterdam, pp. 231-240.

Åman, P. and Nordkvist, E. 1983. Chemical composition and *in vitro* degradability of botanical fractions of cereal straw. Swedish J. Agric. Res., 13, 61-67.

Changes in the physical/chemical structures of cell walls during growth and degradation

F.M. Engels.
Department of Plant Cytology and Morphology, Agricultural University, Wageningen, The Netherlands.

Digestibility of cell walls from different plant species has been of great interest during the last decades. This is not surprisingly since the chemical composition of the cell walls excists of polymers of hexoses and pentoses, known as hemicellulose, cellulose and pectines, which can serve as a carbon and energy source for microorganisms. Digestion experiments in vivo, in vitro, and in sacco with rumen liquor showed great differences in digestibility of the plant cell walls. These differences were found to be linked to the lignin concentration. This lignin, another cell wall polymer, composed of phenolic derivatives, is bound to the cellulose-hemicellulose moiety of the cell wall and acts as a barrier in the enzymatic degradation by bacteria or fungi. Only a few of these microorganisms are able to break down lignins under aerobic or anaerobic conditions.

In order to investigate the digestion of cell walls the plant material is milled, passing a 1 mm sieve, and incubated with rumen liquor. Chemical analysis are performed according to the method of Goering and Van Soest (1970). In most experiments it is assumed that the plant particles are all the same in chemical composition, size and in their chance to be digested by the microorganisms during incubation. Investigating the milled plant material by lightmicroscopy (LM), it proved that material is very heterogeneous with respect to particle size, simple colour reactions on cell wall components and thickness of the cell walls of the multicellular particles. The same can be observed studying the faeces. In both cases it is still unclear which tissue has been digested and why are there still these multicellular particles.

Akin (1982) introduced the section to slide technique which made it possible to investigate the cell wall degradation on the cell wall and the tissue level.

This technique uses thin sections which are sticked on double adhesive cellotape on a glass slide and subsequently are incubated in rumen liquor. Afterwards, a careful investigation of the plant cell wall and plant tissue can be carried out by means of LM or electronmicroscopical (EM) techniques. It was found that some tissues in the section (e.g. the phloeem and paren- chyma) are rapidly degraded whereas other tissues (e.g. vascular bundles) are not attacked at all. LM staining methods on lignin showed that tissues containing lignin differ considerably in their phenolic components, e.g. the vascular bundle is red-purple coloured with the phloroglucinol-HCl colour reaction, indicating the presence of coniferyl, while sklerenchyma is rose-red coloured with the chlorite-sulfite colour reaction, indicating the presence of syringyl.

In 1984 we started to investigate the digestion of barley straw using novel LM and EM techniques. In the material and methods additionally short abbrevia- tions are indicated which will be used in the text.

Materials and methods

The LM and EM techniques used are extensively described by Engels and Brice (1985).

The abbreviations used in the text refer to specific procedures in that paper.

EM-Sc: scanning electron microscopy (SEM).

EM-Re: transmission electron microscopy (TEM) (1) replica.

EM-Ag: PATAg method for vicinal glycol detection.

EM-Mn: $KMnO_4$ staining for phenolic components.

LM-C: chlorine-sulfite test for lignin.

LN-P: phloroglucinol-test for lignin; 1% phloroglucinol in 20% (w/v) HCl.

In EM-Sc prepared sections which were incubated for some time with rumen liquor, the bacteria were found nearly exclusively on the cut-end edges of the cell walls. It was observed that fungi were only very limited present. The inner cell walls were nearly free of bacteria (fig. 1).

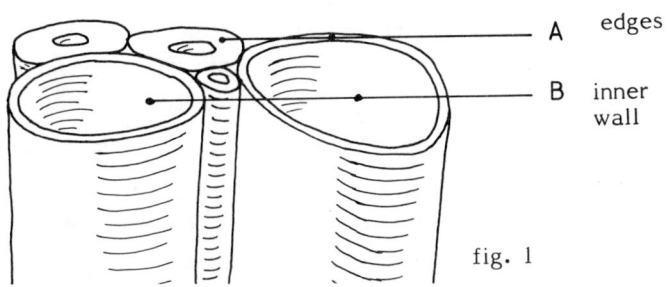

fig. 1

A close examination of the surface of the inner cell walls by means of EM-Re showed that some bacteria were present on a surface which appeared very granular in structure. This granular surface was already established in numerous woody tissues and was named the warty layer (Liese 1963). The surface proved to be very resistant to bacterial attack (Schmidt and Liese 1982) and to strong chemical solvents (Wardrob et al 1959). Only some fungi (Basidiomycetes) have been found to be able to break down this layer under aerobic conditions. In our experiments it was shown that prolonged incubation with rumen liquor (up to 72 h) the warty layer remained undamaged, indicating that bacteria are unable to digest this layer under anaerobic conditions. In EM-Ag preparations we observed that bacteria are closely associated with a cell wall cover which, with this technique appeared to consist of a fine-rough dark stained granular layer (= warty layer) with no evidence of any degradation. It was also observed that bacteria can be located between two dark layers covering the cell walls. Cell wall digestion was very effective in highly lignified cell walls. The cell wall can only be digested if the bacteria are contacting directly the cell wall. In fig. 2 it is illustrated that a cell wall of cell A, can only be digested after the removal of the surrounding cells c.q. cell walls. This means that highly digestible material has not been digested due to the absence of bacteria. This material is generally determined as part of the indigestible fraction.

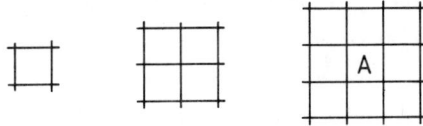

fig. 2

The use of $KMnO_4$ and PaTAg as histochemical stains in EM enables the visualization of different cell wall components. In EM observations, cellulose
will appear as very fine blackdotted lines on an amorphous fine granular
(=hemicellulose) background after PaTAg staining. The lignin stained by
$KMnO_4$ shows a fibrillar skeleton due to the cellulose fibrils which are embed-
ded in the lignin and thus the lignin will also be detecteble as fine lines.
A carefull examination of EM-Ag cell wall preparations resulted in at least
four modes of bacterial action on the cell wall. These are illustrated in
fig. 3.

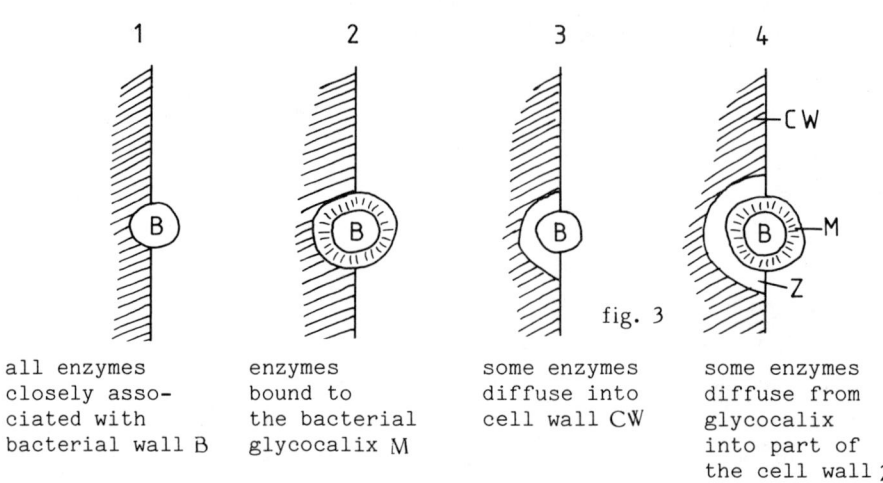

fig. 3

1	2	3	4
all enzymes closely asso-ciated with bacterial wall B	enzymes bound to the bacterial glycocalix M	some enzymes diffuse into cell wall CW	some enzymes diffuse from glycocalix into part of the cell wall Z

It seems that the bacterial enzymes are only active very close to the bacterial surface. Since cellulose and hemicellulose can be distinguished with
the EM after staining with PaTAg this method can be used to discriminate
between cellulolytic, hemicellulolytic and cello-hemi-cellulolytic bacteria
by analysing the pattern of lysis. It was very surprising to observe a white
zone in the cell wall in EM-Mn prepared sections suggesting that the lignin
was already removed over a short distance. Also direct lysis of the cell
wall was observed. We suppose that the bacterial enzymes split off the lignin
fraction from the carbohydrate without further fermentation. Phloroglucinol
staining performed on digested material showed weak colour reaction while
after prolonged digestion the staining was negative.
The 1 mm plant particles are very heterogeneous in composition and size.
Moreover the EM-Ag and EM-Mn staining indicated that cell walls are not
homogeneous. There are regions or layers in every cell wall that vary considerably in their cellulose, hemicellulose and lignin content. However,
chemical analysis of two morphological different types of cell walls may

result in the same chemical composition but their structure and texture
may be different, and vice versa. Chemical treatment (e.g. NaOH extraction)
can increase digestibility due to the removal of lignins but the enormous
swelling of the cell walls has often been overlooked and has not been considered as a possible explanation of the increase in digestibility.
Some remarks have to be made on the techniques used to investigate the warty
layer. The EM-Sc will not always allow the visualization of the warty layer
because its resolution is too low. This resolution is determined by the
electron spot diameter and the distance in the cell wall from which the
secundary electrons are emitted. This problem is not met in the EM-Re technique. The resolution that can be reached depends to the platina grains and
the better the vacuum during the evaporation of the platina the smaller
the grains will be. In practice, cellulose elementary fibrils can be investigated (40-30 Å) by means of the EM-Re technique. In fig. 4 part of a cell
wall has been digested by bacteria and cellulose fibrils can be observed.

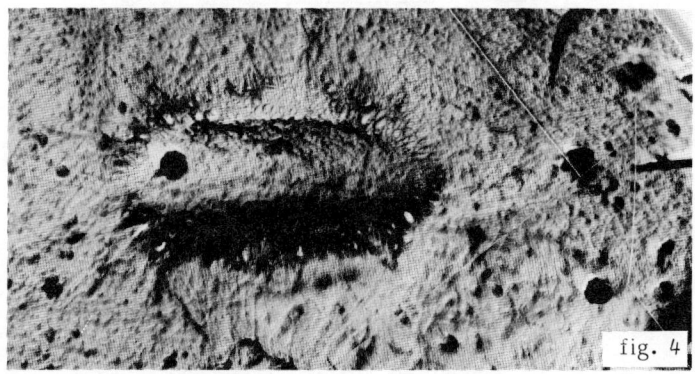

fig. 4

The use of Mn to investigate lignified cell walls may result in a mis-inter-
pretation because the KMnO$_4$ did not penetrate very well into the cell wall
during fixation. Therefore, in our experiments we determined the fixation
time by a delignification treatment followed by a phoroglucinol test in
a parallel experiment. When this test became negative fixation was stopped.
Maize plants were grown either at 18/12°C or at 30/24°C (12/12 h day/night
rithm). The 30/24 plants were digestible for 40% after 48 h in vitro digestion
while the 18/12 plants were digestible for 70%, when harvested during pollina-
tion (Struik et al 1985).
A cell wall analysis carried out according to Goering and Van Soest showed
a nearly twofold higher lignin content in 30/24 plants (table 1).

| 18/12 | cellulose 29 | hemicellulose 21.2 | lignin 2.8 | |
| 30/24 | cellulose 31 | hemicellulose 21.0 | lignin 5.0 | in %% |

table I

A LM-P colour reaction on stem sections proved to be the more intensive
in the 18/12 plants while a LM-C colour reaction proved to be more intensive
in the 30/24 plants. This would mean that 30/24 material should be more
digestable than 18/12 material since syringyl tissues have been found to
be better digestable. In vitro digestion showed the opposite results. It
can be concluded that plants grown under different temperature regimes posses
a different lignin composition in their comparable cell walls.

The rate of in vitro digestion of 18/12 and 30/24 milled 1 mm maize stems
was determinated (fig. 5). It is suggested that both maize types are composed
of two different types of tissues. One which is digested rapidly and an
other which is digested slowly. According to the breakpoint in fig. 5 after
14 h of in vitro digestion of sections only the tissues with a lower digestibility should remain.

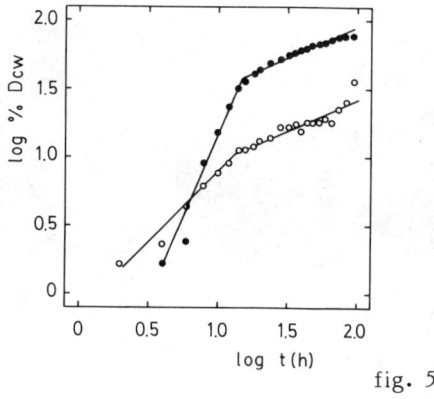

fig. 5

We did not observe this effect at 14 h of incubation but at 20 h. The difference in time is because the 1 mm particles can be digested from all directions while the sections can only be attacked from one side in the section
to slide technique. Considerable differences in the degree of digestion
between similar tissues in the maize sections in EM-Sc preparations (fig. 6)
were also observed.

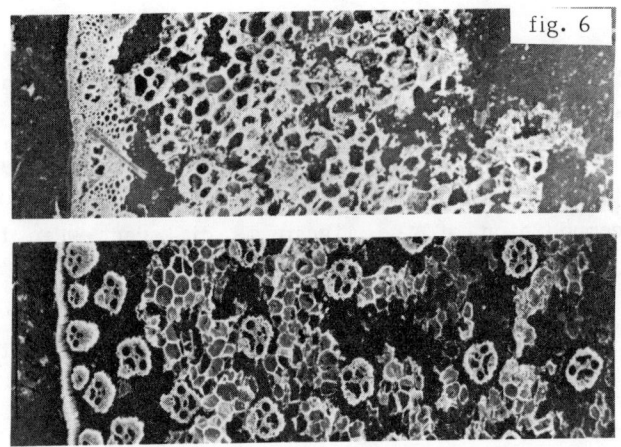

fig. 6

The warty layer can be easily overlooked in LM because the layer is often very thin and has a same birefringuence as the cell wall (Liese 1963). We treated the cell walls in such a way that a fine granular layer became visible. This method was applied on barley straw and on several grasses. It was found that tissues showing clearly the granular layer (LM) were not digested within 24 h. If the granular layer was not detectable the tissue was digested within 8 h. In maize the same observations were made. The differences in the degree of digestion between similar tissues of 18/12 and 30/24 maize sections were correlated with the absence or presence of this granular layer. At this moment we suppose the LM granular layer is identical with the EM warty layer which is supported by other experiments. Whenever very thin cell walls (especially the parenchyma in 30/24 maize material) were not digested it was found that bacteria can not penetrate between the two adjacent warty layers at the cut end edges of the cell walls (fig. 7).

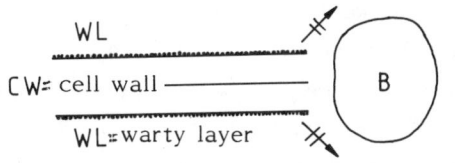

fig. 7

When plants grow older, digestibility decreases (Struik 1985) while the lignin content of the cell walls increases. We therefore analysed 1M NaOH extractable phenolic acids at different times during the development of the plants. After the extraction the cell wall residues were no longer stainable with LM-P and LM-C techniques. Barley and wheat straw both showed a weak colouring after the same extraction. The penolic acids were analysed on a GLC capillar collumn. Coumaric and ferulic were found as the main phenolic acids. This research is still pending.

Conclusion

The results obtained from the investigations on barley and wheat straw, maize and grasses with LM and EM techniques are recent and priliminar results. There are at least two more factors which determine the digestibility of cell walls, that is the presence of a warty layer and the thickness of the cell wall. The composition of the lignin is different when the plants are grown under different temperature regimes. Within one section there are tissues with a different lignin composition. The differences in staining after NaOH treatment of cell walls shows a plant specific lignin composition. As the results are presented here further investigations are required to trace the origin and the chemical composition of the warty layer.

The present paper is based on very recent work on cell wall structures and degradation in which Dr. J.W. Cone, Dr. R.E. Brice and the author closely cooperated.

Refereces

Aken, D.E. (1982). Section to slide technique for study of forage anatomy and digestion. Crop Sci. 22: 444-446.

Engels, F.M. and R.E. Brice (1985). A barrier covering lignified cell walls of barley straw that restricts access by rumen microorganisms. Curr. Microbiol. 12: 217-224.

Goering, H.K. and P.J. van Soest (1970). Forage fiber analyses (apparatus, reagents, procedures and some applications). In: Agriculture handbook no 379. Washington DC: Agricultural Research Service, US Department of Agriculture.

Liese, W. (1963). Tertiary wall and warty layer in wood cells. J. Polynez Sci. (C) (Polynez symposium no 2), pp. 213-219.

Schmidt, O.R. and W. Liese (1982). Bacterial decomposition of woody cell walls. Int. J. Wood Preservation 2(1): 13-19.

Struik, P.C., B. Deinum and I.P.M. Hoefsloot (1985). Effect of temperature during different stages of development on growth and digestibility of forage maize (Zea mays L.). Neth. J. Agric. Sci. 33(4) (in press).

Wardrob, A.B., W. Liese and G.W. Davis (1959). The nature of the wart structure in conifer tracheids. Holzforschung 13(4): 115-120.

OPTIMAL DEGRADATION OF LIGNOCELLULOSIC FEEDS BY RUMINANTS AND IN VITRO DIGESTIBILITY TESTS

J.M. van der Meer[1] and A.J.H. van Es[1,2]

[1]Institute for Livestock Feeding and Nutrition Research (IVVO), P.O. Box 160, 8200 AD Lelystad, The Netherlands.

[2]Department of Animal Physiology, Agric. Univ., Haarweg 10, 6709 PJ Wageningen, The Netherlands.

ABSTRACT

Ruminants, thanks to their symbiosis with micro-organisms in forestomachs and large intestine, can use lignocellulosic plant materials as feed. Voluntary intake and digestibility of these feeds mainly determine their feeding value for this kind of livestock.
Nutrient requirements for maintenance and animal production are described and the effects of degradability on intake and availability for resorption are discussed. The structure of the plant cellwalls together with circumstances for bacterial activity determine the degradability and degradation rate in the animal. Degradation processes and rate of lignocellulosic materials can be studied in sacco and in vitro tests. Finally the use of lignocellulosic feeds for optimal production is discussed.

INTRODUCTION

Lignocellulosic plant materials are not suited as food for man or farm animals such as pigs and poultry because these do not possess the enzymes necessary for their digestion. Ruminants, thanks to their symbiosis with micro-organisms in forestomachs and large intestine, can use such materials as a feed. Voluntary intake and digestibility of these feeds mainly determine their feeding value for this kind of livestock.

NUTRIENT REQUIREMENTS AND VOLUNTARY INTAKE

Nutrient requirements

It is widely accepted (A.R.C., 1980, McDonald et al., 1981) that in general voluntary feed intake of ruminants increases with production level, i.e. with increasing nutrient requirements. Appetite in growing ruminants, relative to body weight decreases with age, i.e. when growth rate slows down. Lactating cattle eat more than non-lactating ones although in cows with a high potential for milk yield it often takes some time after calving before intake has risen sufficiently.

Therefore it can also be expected that voluntary intake will be reduced in a producing animal when the production cannot reach its maximum because an essential nutrient is in short supply. This is probably the explanation why growing or lactating ruminants increase their intake when a deficiency of absorbed amino acids is corrected, e.g. by means of by-pass protein.

The size of the forestomachs in relation to body weight, their possibility to enlarge after a meal and rate of passage of the digesta mainly determine if sufficient nutrients will be absorbed from the ration offered. Rate of passage is certainly not only dependent of the diet, also animal factors play a part, e.g. rate of intake, intensity of chewing, saliva production, rate of rumination, rumen mobility and size of the reticulo-omasal orifice. It is not easy to take account of all these factors in a selection program, moreover because the selection has to be done under more or less similar circumstances as occurring in practical farming.

Chesson and Orskov (1984) mention that there are many indications suggesting that because of the use of high-quality feeds European and North American cattle have been selected in the course of time against maximal roughage utilization while for cattle in the (sub)tropics the reverse holds true as a result of their low-quality feed supply. However, in the former animals the high milk yields also necessitated a high total intake.

Among ruminant species and probably also within species forestomach volume and rumination rate increases with body weight -kg-, maintenance requirement only with metabolic weight -$kg^{\frac{3}{4}}$- (Van Soest, 1982). Thus large animals need in the same time interval per unit of forestomach volume a smaller quantity of fermentation products for maintenance. So, they can afford to eat feeds that require more time of fermentation being a fairly slow process. Small ruminants, indeed, have a different feed selection behaviour and prefer plant parts with higher cell contents and less lignified cellwall. Their smaller mouth enables them to eat far more selectively than large animals.

Most lignocellulosic feeds have low ME contents. Just to cover their maintenance requirements steers of 400 kg have already to eat 6.4 kg dry matter of a feed like straw containing 7 MJ per kg dry matter, more than 1.5% of their body weight. For such a feed it is close to their maximum. From this example it will be clear that when lignocellulosic feeds have to be used, everything possible should be done to obtain maximal intake, by technological treatment and by optimizing the conditions for intake in the animal. Only in this way will it be possible that acceptable rates of pro-

duction will be reached. It appears probable that large animals have some
advantage in this respect, especially if selective feed intake is undesired.

Effects on intake by ruminants

Intake capacity of ruminants of low quality feeds, often the main feed
in tropical and subtropical areas, depends to a large extent on rate of feed
degradation in the mouth and in the forestomachs (Nicholson, 1984; Minson,
1985; Van Es, 1985). Particles longer than 0.1 cm are selectively withheld
for a longer time in the forestomachs of sheep, goats and camels (Heller et
al., 1985), so reducing the possibility for renewed intake of feed. Also in
cattle, despite the fact that the reticulo-omasal orifice has a largest diameter of 2 cm, particles of 0.2 cm and more are retained for a longer time
in the forestomachs than smaller ones (Welch, 1982). On the other hand this
has of course the advantage that the cellulolytic microbes get more time to
attack the cellwall carbohydrates resulting in a welcome volatile fatty acid
(VFA) supply for the host animal. Also distension of the forestomachs adds to
higher voluntary intake by the animal and to longer opportunity for microbial
degradation of the feed (Von Engelhardt, 1985).

The reduction of particle size is done by chewing, rumination and microbial degradation. Thus, it is essential that the microbes in the forestomachs
meet a good environment, the more so as their growth results in an important,
often even the main protein supply to the host animal. Microbial activity and
growth in the forestomachs can be reduced by lack of N, S and minerals, by
low pH -e.g. after eating large amounts of easily fermentable feeds in a
short time- and by low degradability of the feed due to severe lignification,
high tannin or silica content, etc. In fact, microbial degradation proceeds
best when degradable feed, N, S and minerals during the 24 h of the day are
always present in sufficient quantities. Thus for instance urea additions
aiming at increasing the N content of a low N diet should preferably be given
in many small portions during the day or well-mixed with the feed (Kellaway
and Leibholz, 1983). Browse also is a valuable source of N in cases of low N
supply (Otsyina and McKell, 1985).

Absorption of nutrients from the gastrointestinal tract of ruminants

While maturing, all plants tend to increase the amount of (hemi)cellulo-
ses, lignins and sometimes tannins and silicae in their stalks and leaves, so
digestibility will decrease. Moreover, the protein content will fall during
seed formation when this occurs uninterruptedly (Nicholson, 1984). Maturation

in the tropics proceeds more rapidly and to a greater extent (Van Soest, 1978).

Celluloses and hemicelluloses are β-linked carbohydrates for which animals (man included) do not themselves have the necessary enzymes for degradation. Through the anaerobic symbiosis with microbes in forestomachs and/or large intestines that do possess the necessary cellulases and hemi-cellulases, still such cellulosic material can be used as feed. However, the symbiosis involves an energy loss: the microbes use part of the energy of the (hemi)cellulose for themselves.

Fortunately, the time that the microbes can do their work is far too short for complete conversion of (hemi)cellulose into CO_2, H_2O, CH_4 and heat (Bryant, 1979). Instead, the degradation is partial, some 10% of the energy of the degraded material becomes energy in CH_4, small parts, variable in size, a.o. depending on rate of passage, become heat and energy in microbial matter and the major part is converted into VFA.

This microbial fermentation is not restricted to (hemi)cellulose, all carbohydrates and proteins entering the fermentation vessels (forestomachs, hind gut) are attacked. There is clearly a preference for α-carbohydrates and proteins and for easily fermentable β-carbohydrates i.e. those compounds which are not too much enclosed by or infiltrated with lignins, tannins, silica, etc.

When much easily fermentable material is ingested by the animal at a time pH of the fluid in the forestomachs may fall as a result of high production of VFA (De Visser, 1982). In such cases absorption of VFA is not rapid enough. Furthermore, the buffering capacity of the rumen fluid is low because such material needs little chewing so that with the saliva little bicarbonate enters the forestomachs (Counotte, 1981). Some compensation for the reduced fermentation of plant cellwall in the forestomachs may occur in the large intestine (Ben Ghedalia and Miron, 1984). However, it is the general opinion that the circumstances for microbial fermentation in the forestomachs are better than in the large intestine. Moreover, the host animal does not bene-fit from absorption of amino acids from protein synthesized by microbes in the large intestine because such protein is either excreted or degraded to ammonia. Some of the ammonia may be absorbed, converted into urea and used as N-source in the forestomachs if there exists an N-shortage.

The rate of fermentation depends on the properties of the environment in which it takes place. It is reduced when the microbes cannot obtain sufficient quantities of N and, probably, S and some peptides or amino acids. The need for N is related to the fermentability of the plant material. When this is low, per unit of time little energy becomes available so that also only small amounts of N are needed. A low pH, e.g. caused by a high intake of easily fermentable feed, as mentioned above, has a negative influence. Most cellulolytic bacteria as well as methane bacteria rapidly reduce their activity at a pH of 6.5-6.0 and lower (Mould and Ørskov, 1983). It is accompanied by an increase of the propionic acid to acetic acid ratio of the VFA produced.

Structural carbohydrates are almost insoluble. So micro-organisms must find other ways to reach their substrate. Protozoën have something like an oral cavity through which fibre particles can be collected and digested (Coleman et al., 1976).
In the predominant cellulolytic rumen bacteria adhesion is a prerequisite to cellulolysis because the cellulase enzymes are bound to the cell surface although some strains also release vesicle bound enzymes (Forsberg, 1981). All the cellulolytic rumen microorganisms can also attack, to some degree, the hemicelluloses of the plant cellwall. In some species cellulase and xylanase activity is associated with the same enzyme complex. Other bacteria may attack hemicelluloses but not cellulose. All those bacteria act in a consortium resulting in extensive digestion of the plant cellwalls.
There is a time lag after feed ingestion by the ruminant before the degradation of structural carbohydrates proceeds well because of the adhesion and the necessary bacterial adaptation to the ration.

Entry of microbes in the plant material is by preference via damaged surfaces and natural openings such as stomata (Chesson and Ørskov, 1984). Cellulolytic microbes must first attach to feed particles for which both time and an easy energy source are needed. In mature materials such easily available energy -sugars, starch, proteins- is scarce. Moreover, when meals are separated by long intervals, microbes may decrease in number because of lack of N and/or energy and the microbial population may change. Both Chesson and Ørskov (1984) and Preston and Leng (1984) mention the importance of a lasting large pool of microbes in the rumen for digestion of low-quality feeds.

It is interesting to know what will happen if microbes in the forestomachs are restricted in their growth by lack of N. Will they, in case of sufficient supply of easily available carbohydrate, continue to produce VFA while reducing their ATP production by further uncoupling and using some nutrients and ATP for fat synthesis? When such a situation continues for a longer period it is probable that some microbial species increase in number whereas other species decrease because they are more sensitive to sub-optimal circumstances.

Reduced rate of fermentation not only means lower supply of VFA and absorbable amino acids of microbial origin for the host animal. It also decreases rate of passage of the feed and therefore intake.

DEGRADATION OF LIGNOCELLULOSIC MATERIAL IN "IN VITRO" STUDIES

Many materials rich in lignocellulosics have been analysed by chemical methods such as detergent extractions (Van Soest and Wine, 1967) but the results of such methods often only insufficiently predict the degradability by bacteria. Contents of hemicelluloses, celluloses, pectins and lignins do not show the degree of polymerisation, extent of cristallinity or encrustment of the polysaccharides by the polyphenols. Also within the cellwall distribution of cellwall constituents is not uniform which may influence the ways of microbial digestion.

Phloem is highly digestible while the parenchymal cells are slightly slower digested. The remaining vascular bundles together with epidermus are (almost) not degraded (Brice, 1985). Chemical (sodium hydroxide, lime, ammonia, sulphur dioxide) or physical (soaking, milling, steaming) treatments considerably change degradability and so often increase intake which however is often poorly reflected in changes in the chemical composition.

Recent physical analysis explained much better what is happening to cellwalls on treatment. From electron microscopic studies and pyrolytic mass spectrometry it became clear that on ammonia treatment the cellwall layers are released from each other and on treatment with sodium hydroxide these layers are disturbed, in both cases without important changes in chemical composition. By treatment also the ADF content does not change but the digestibility of ADF is influenced (Solaiman, 1979; Herrera, 1983).

For many years already, in vitro digestibility techniques have been used to predict the digestibility of roughages, wastes and byproducts for ruminants. As sais above the content of digestible organic matter is one of the two most important factors determining a feed's feeding value. In vitro procedures incubate feed samples first with buffered rumen juice and then with pepsin/HCl measuring the organic matter disappearance (Tilley and Terry, 1963). When standard samples with known in vivo values of materials similar to the unknown samples are included the organic matter- or cellwall digestibility for ruminants can be predicted with reasonable precision (Van Es and Van der Meer, 1980).

Feed-animal and feed-method interactions occur. Passage rate in the animal and digestion rate in vitro may differ considerably so that the interpretation of the results cannot be straight forward. Without an appropriate correction the improvement due to e.g. a chemical treatment is overestimated and differences found between ammonia and sodium hydroxide do not reflect the in vivo situation.

There is a tendency in in vitro studies in replacement of rumen juice by fungal cellulase enzyme preparations. The degradation of lignocellulosic materials by these enzymes is often only 50% of the physiological degradation (Van der Meer, 1981), so standard samples with known in vivo digestibility are even more needed to translate cellulase degradation results to physiological circumstances.

Besides digestibility, voluntary intake is the other major factor in energy availability for the ruminant. Degradability, often combined with a low degradation rate, results in a low passage rate. It has been described already how circumstances for digestion can be optimalised, which will mainly shorten the lag time and speed up degradation rate.
Degradation rate can be studied in nylon bags incubated in the rumen of ruminants fitted with a large rumen cannula. A number of bags containing the same lignocellulosic feed are introduced at the same time and removed from the rumen at different times. On the assumption that the digestion process is undisturbed the degradation follows first order kinetics (Smith et al., 1971) sothat degradation rate, undegradable residue and the lag time can be derived.

Also in vitro procedures can be applied to study organic matter or cellwall degradability. Comparison studies showed similar results for cellwall degradation in vitro and in sacco (Fig. 1). Only lag time was different due to the relatively small number of bacteria introduced at the start of the in vitro test and the loss on particles at the start of the in sacco experiment (Cabrera and Van der Meer, 1986). In vitro tests are more suited to study the effects of additions to less degradable feeds aiming at improving rate and extent of degradation.

Part of the effect of feeding level can be studied using rumen juice from animals maintained at different feeding levels but in the in vivo situation feeding level also increases rate of passage. As disappearance is measured soluble products like lignosulphonates and molasses cannot be studied by in sacco or traditional in vitro methods.

The "Gasbildungstest" is a procedure in which the gas (CO_2 + CH_4) produced on microbial degradation is measured in calibrated piston syringes (Menke and Raab, 1979; Fig. 2) which also gives some information on rate of degradation.

Traditional in sacco, in vitro and the "Gasbildungstest" procedures produce supplementary information, i.e. OM- or NDF-disappearance and gas production in time (Fig. 3).

All these laboratory procedures can be used to predict both degradability and degradability rate if sufficient corrections can be made by the use of standard materials with known figures obtained in animal trials.

OPTIMAL USE OF LIGNOCELLULOSIC FEEDS BY RUMINANTS

The use of lignocellulosic materials as the main feed

From what has been said above it will be clear that everything possible should be done to maximize the rate of fermentation. Of course, the measures to be taken should be economically feasible. While deciding on this, all costs and all benefits -higher intake, higher nutrient supply, higher production to maintenance ratio- should be taken into account.

Giving slight additions to the lignocellulose diet so that it meets the requirements of the microbes better -some easily fermentable carbohydrate for their rapid attachment to plant particles, so much NPN as needed in view of the rate of degradation of (hemi)cellulose and preferably some protein, some S and minerals if there is a possible deficiency for the microbes-, speeds up rate of fermentation. Diet and such additions preferably should be available at every moment during the 24 hours of the day and in the mixed form; anyway, excessive intake of additions should be prevented. If it is feared that absorption of amino acids is too low it should be tried if giving more protein, by preference of low rumen degradability but good intestinal digestibility, improves intake and performance.

Intake can be further stimulated by treatment of the lignocellulosic material or by offering large quantities of such material if heterogeneous to permit selective uptake of the better parts. However, the advantages of these two procedures should exceed their costs and possible negative effects -grinding the feed usually results in a higher intake, but also in a lower digestibility of the cellwalls.

Heat stress resulting in lower roughage intake should be lessened by all means. The following measures should be considered: parturition in such parts of the year that highest gain or milk yield falls in the coolest months, feeding during or just before the cooler part of the day, short walking distances to watering and milking place; preventing insects, dogs, farmhands to upset the animals; improving the quality of the diet during warm and humid periods.

The use of lignocellulosic material as part of the ration

When the lignocellulosic material is of so low quality that despite additions, treatment and other measures mentioned above, it gives too low intake and too low absorption of nutrients, its use in a mixed ration with feed of better quality might be possible. Of course, also lignocellulosic feeds of better quality can be used in such a way. To obtain most profit from such feeds under these circumstances, attention should be paid to way of feeding as well as type of the better-quality feed. Offering the whole diet in a well-mixed form prevents the animals to select the better quality feed. This better feed should not ferment too rapidly, because this would lead to lower pH and unfavourable conditions for fermentation of the low quality feed. Thus the better feed should contain mainly not too heavily lignified (hemi)cellulose and some, preferably slowly, degrading α-carbohydrate (from corn, sorghum, rice rather than wheat or barley).

Such mixed rations will not meet the requirements of high-yielding dairy cows (Van Es, 1978, 1980) as intake and absorption of nutrients will be too low because of the slow degradation of low quality feed which also leads to high fill of the gastrointestinal tract.

The mixed rations might be fortified with fat (Van der Honing, 1983). For the microbes in forestomachs and hindgut fat has hardly any value as an energy source because it has a very high H to O ratio. When the fat particles cover the lignocellulosic material, this will less easily be fermented. Moreover, the fatty acids resulting from lipolysis of the fat in the forestomachs especially the unsaturated ones have some antimicrobial activity. For the animal fat is an excellent energy source, however, unsaturated fat, probably through their effect on fermentation, may lower milk fat content. Fats that have not been preserved well may depress the animal's appetite.

REFERENCES
A.R.C. (Agricultural Research Council) 1980. The nutrient requirements of ruminant livestock. Unwin, Surrey, 351 pp.
Ben-Ghedalia, D. and Miron, I. 1981. Effect of sodium hydroxide, ozone and sulphur dioxide on the composition of in vitro digestibility of wheat

straw. J. Sci. Food Agric., 32, 224-228.

Boucqué, Ch. V., Fiems, L.O., Cottijn, B.C. and Buysse, F.X. 1984. Feeding value of some wastes and by-products of plant origin and their use by beef cattle. In: Animals as waste converters (Ketelaars, E.H. and Boer Iwema, S., eds.), Pudoc, Wageningen, p. 54-61.

Brice, R.E. et al. 1985. Ammonia treatment of barley straw. Characterisation of untreated and ammonia treated straw by light and electron microscopy. (to be published)

Bryant, M.P. 1979. Microbial methane production - theoretical aspects. J. Anim. Sci., 48, 193-201.

Butterworth, M.H. 1984. Animals in relation to land use. In: World animal science A2 (Nestel, B., ed.), Elsevier Amsterdam, p. 15-32.

Chesson, A. and Ørskov, E.R. 1984. Microbial degradation in the digestive tract. In: Straw and other fibrous by-products as feed (Sundstøl, F. and Owen, E., eds.), Elsevier, Amsterdam, p. 340-372.

Coleman, G.S., Laurie, J.I., Bailey, J.E. and Holdgate, S.A. 1976. The cultivation of cellulolytic protozoa isolated from the rumen. J. of Microbiology, 95, 144-150.

Engelhardt, W. von. 1985. Contribution to ACSAD/AOAD conference on animal production in Arid Zones, Damascus.

Es, A.J.H. van. 1978. Feed evaluation for ruminants. Livest. Prod. Sci., 5, 331-345.

Es, A.J.H. van. 1980. Net requirements for maintenance as dependent on weight, feeding level, sex and genotype, estimated from balance trials. Ann. Zootech., 29, 73-84.

Es, A.J.H. van and Meer, J.M. van der. 1980. Methods of analysis for predicting the energy and protein value of feeds for farm animals. In: Proc. Workshop on Methodology of analysis of feedingstuffs for ruminants (Es, A.J.H. van and Meer, J.M. van der, eds.), Pudoc, Wageningen, p. 59-66.

Forsberg, C.W., Beveridge, T.J. and Hellstrom, A. 1981. Cellulase and xylanase release from Bacteroides succinogenes and its importance in the rumen environment. Applied and Environm. Microb., 21, 886-896.

Gomez Cabrera, A. and Meer, J.M. van der. 1986. Degradation rate of organic matter and NDF fraction of barley straw: Effect of variety and ammonia treatment on in sacco and in vitro degradation.(to be published)

Heller, R. et al. 1985. Physiological aspects of using lignocellulosic materials for animal feed. In: Improved utilisation of lignocellulosic materials in animal feed. OECD, Paris, p. 76.

Herrera-Soldana, R., Church, D.C. and Kellerns, O. 1983. Effect of ammoniation treatment of wheat straw on in vitro and in vivo digestibility. J. Anim. Sci., 56, 939-942.

Honing, Y. van der, et al. 1983. Further studies on the effect of fat supplementation of concentrates fed to lactating dairy cows. Neth. J. Agric. Sci., 31, 27-36.

Kellaway, R.C. and Leibholz, J. 1983. Effects of N supplements on intake and utilization of low quality forages. Wrld. Anim. Rev., 48-33.

Kossila, V.L. 1984. Location and potential feed use. In: Straw and other fibrous by-products as feed (Sundstøl, F. and Owen, E., eds.), Elsevier, Amsterdam, p. 4-21.

McDonald, P., Edwards, R.A. and Greenhalgh, J.F.D. 1981. Animal nutrition, Longman, London, p. 146-186 and p. 211-239.

Meer, J.M. van der. 1981. Prediction of the digestibility of alkali treated straw. In: Utilization of low quality forages in Africa (Kategile, J.A. et al., eds.), Agric. Developm. Rep. 1, Agric. Univ., Aas, Norway.

Menke, K.H. and Raab, L. 1979. The estimation of the digestibility and metabolizable energy content of ruminant feedingstuffs from the gas production when they are incubated with rumen liquor in vitro. J. Agric. Sci. of Camb., 93, 217-222.

Minson, D.J. 1985. Fibre as a limit to tropical animal production. Proc. 3rd AAAP Animal Science Congress, Seoul, May, 1985, vol. 1, p. 108.

Mould, F.L. and Ørskov, E.R. 1983/4. Manipulation of rumen fluid pH and its influence on cellulolysis in sacco, dry matter degradation and the rumen microflora of sheep offered hay or concentrate. Anim. Fd. Sci. Techn., 10, 1-14.

Nicholson, J.W.G. 1984. Digestibility, nutritive value and feed intake. In: Straw and other fibrous by-products as feed. (Sundstøl, F. and Owen, E., eds.) Elsevier, Amsterdam, p. 340-372.

Otsyina, R.M. and McKell, C.M. 1985. Browse in the nutrition of livestock. A. Review. Wrld. Anim. Rev., 53, 33.

Smith, L.W., et al. 1971. In Vitro digestion rate of forage cellwall components. J. Dairy Sci., 54, 71-76.

Soest, P.J. van and Wine, R.H. 1967. Use of detergents in the analysis of fibrous feeds. IV. Determination of plant cellwall constituents. J. Assoc. Offic. Agric. Chem., 46, 819-835.

Soest, P.J. van and Jones, L.H.P. 1968. Effect of silica in forages upon digestibility. J. Dairy Sci., 51, 1644-1648.

Soest, P.J. van, Mertens, O.R. and Deinum, B. 1978. Preharvest factors influencing quality of conserved forage. J. Anim. Sci., 47, 712-720.

Soest, P.J. van. 1981. Limiting factors in plant residues low bio degradability. Agric. Environm. 6, 135-143.

Soest, P.J. van. 1982. Nutritional ecology of the ruminant. Durham and Downey, Portland, 374 pp.

Solaiman, S.G., Horn, G.W. and Owen, F.N. 1979. Ammonium hydroxide treatment on wheat straw. J. Anim. Sci., 49, 802-808.

Tilley, J.M.A. and Terry, R.A. 1963. A two stage technique for the in vitro digestibility of forage crops. J. Brit. Grassl. Soc., 18, 104-111.

Visser, H. de. 1982. Lower starch and sugar content of concentrates for prevention of rumen acidosis and higher energy intake in early lactation. Proc. 12th World Conf. Animal Diseases, p. 415-420.

Welch, J.G. 1982. Rumination, particle size and passage from the rumen. J. Anim. Sci., 54, 885-894.

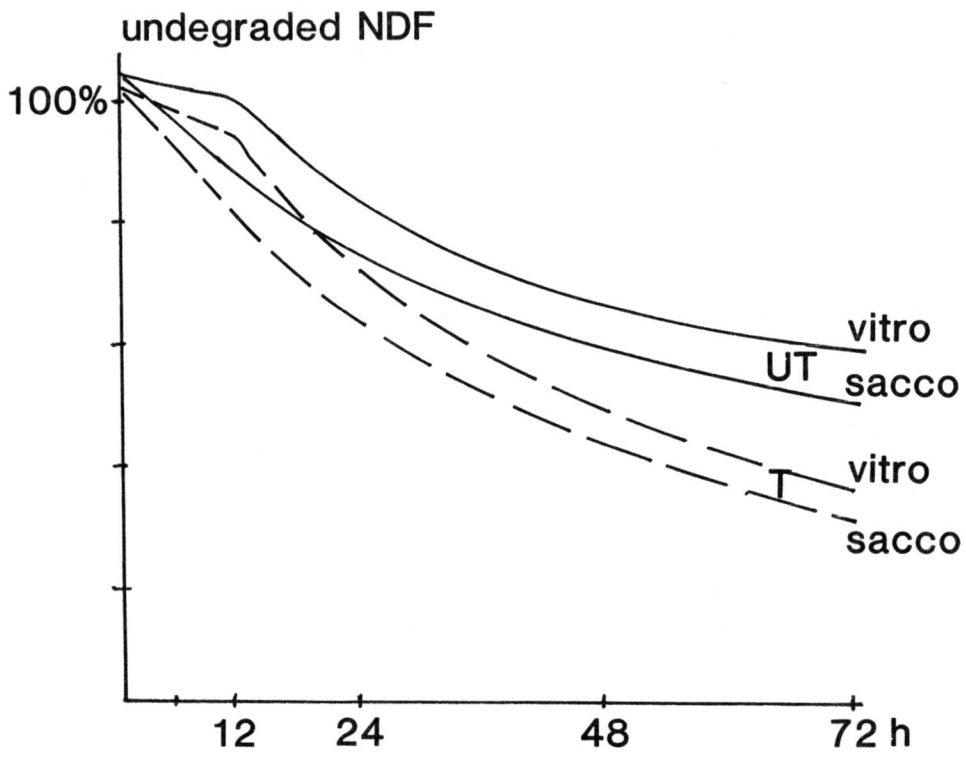

Fig. 1 NDF degradation in time of untreated (T)
and NH₃ treated (UT) barley straw
(Sonja var.) in In Sacco and In Vitro tests.

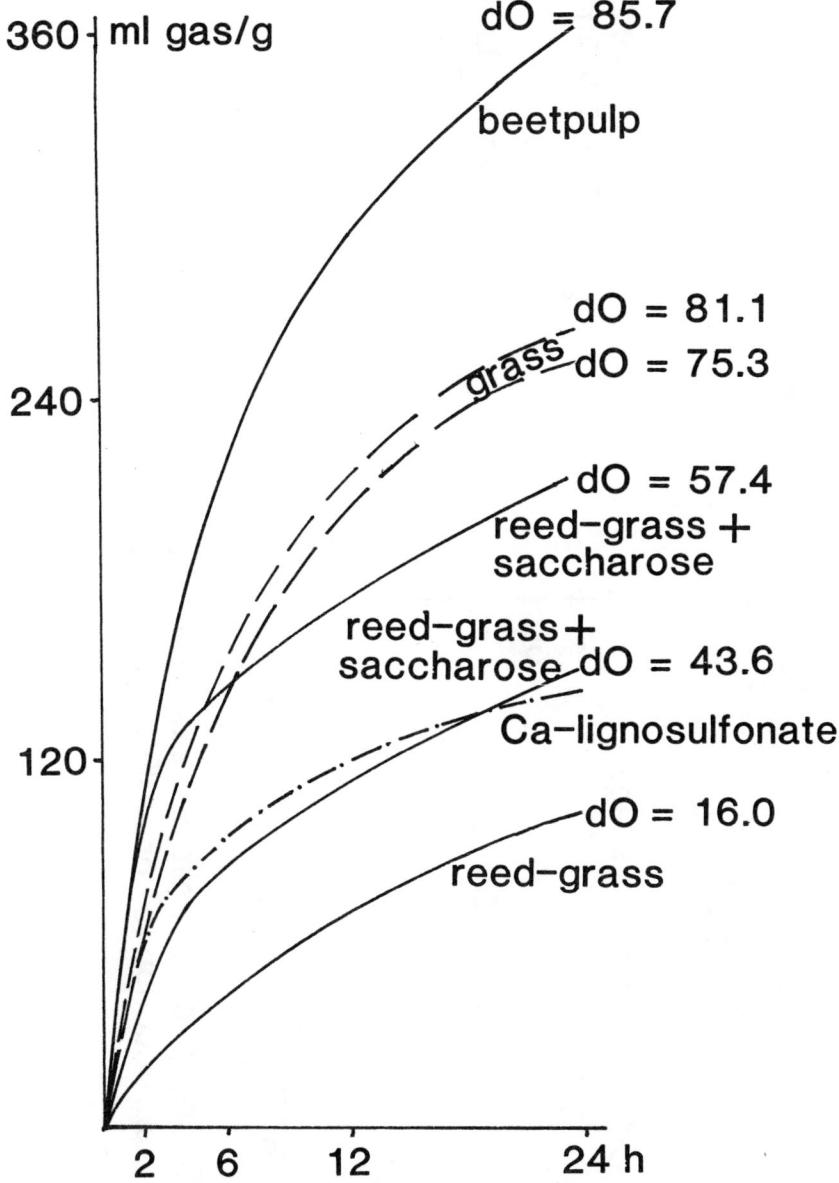

Fig. 2 Gasproduction on incubation with rumen juice

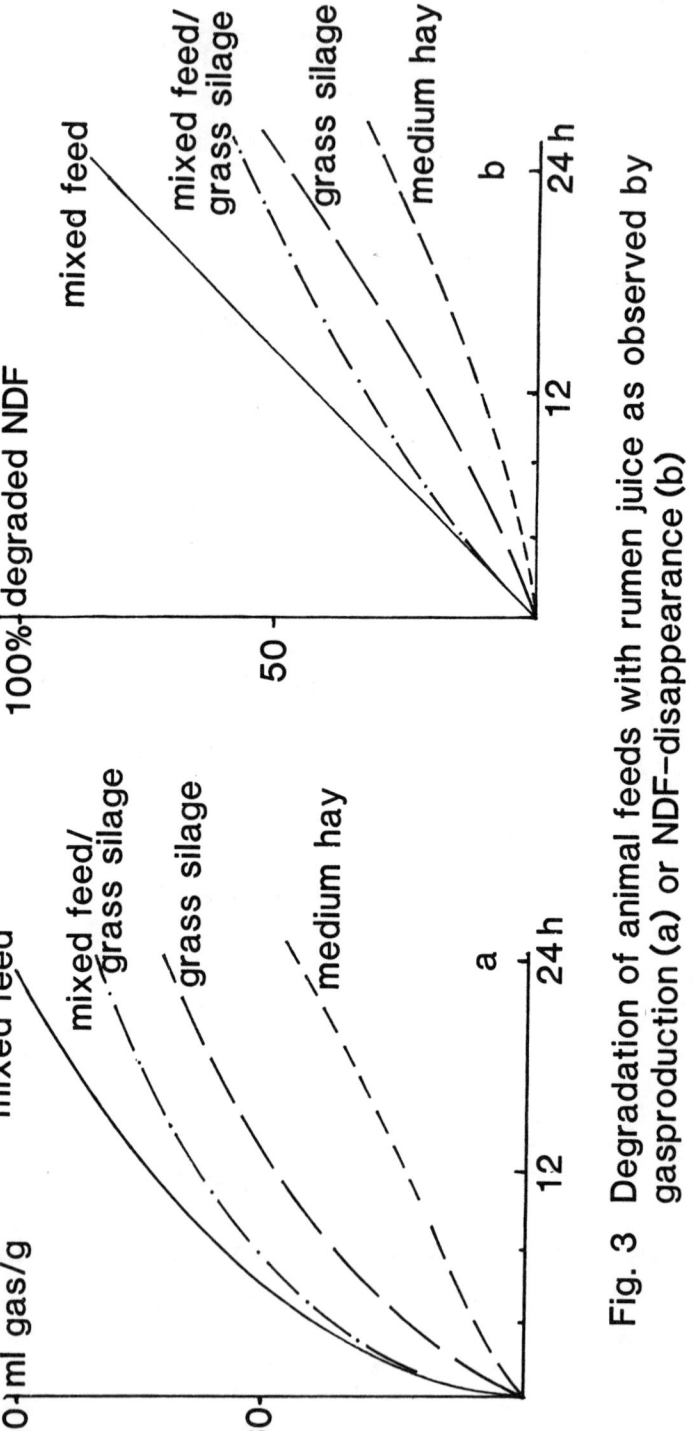

Fig. 3 Degradation of animal feeds with rumen juice as observed by gasproduction (a) or NDF-disappearance (b)

EFFECT OF FUNGAL TREATMENT OF LIGNOCELLULOSICS

ON BIODEGRADABILITY

E. Agosin[1], M.-T. Tollier[2], E. Heckmann[1],
J.-M. Brillouet[3], P. Thivend[4], B. Monties[2] and E. Odier[1]

1/ Laboratoire de Microbiologie, Institut National Agronomique
Paris-Grignon, 16 rue Claude Bernard, 75005 Paris, France.

2/ Laboratoire de Chimie Biologique et de Photophysiologie
végétale, Institut National Agronomique Paris-Grignon,
78850 Thiverval-Grignon, France.

3/ Laboratoire de Biochimie et Technologie des Glucides,
Institut National de la Recherche Agronomique,
rue de la Géraudière, 44072 Nantes Cedex, France.

4/ Laboratoire de la Digestion des Ruminants,
Institut National de la Recherche Agronomique, Theix, 63122 Ceyrat.

ABSTRACT

Delignification in semi-solid cultures using white-rot fungi
Dichomitus squalens and Cyathus stercoreus facilitated 80%
improvement of In Vitro Dry Matter Digestibility (IVDMD) of wheat
straw after 20 days cultivation with 15-20% dry matter loss. Good
correlation was found between lignin degradation and digestibility
while phenolic acid decrease showed no correlation with digestibility
of decayed straw. Water-soluble substances consisting of soluble
sugars as well as lignin degradation products increased in decayed
straw. Extent of digestion of cellulose and xylan was enhanced while
digestion rate of xylan was increased to various degrees depending on
the fungal strain. Improvement of digestibility of decayed straw
appears to result from both solubilization as well as increased
biodegradability of cellulose and xylan in straw.

INTRODUCTION

Agricultural by-products are carbohydrate-rich residues which
represent a potential source of dietary energy for ruminants. However
their feed value is limited by the low polysaccharide degradation
achieved during rumen digestion (Sundstol and Owen, 1984). The
digestibility of these materials is limited by the presence of
lignin which prevents access of hydrolytic enzymes to cellulose and
hemicelluloses. Indeed digestibility of lignocellulosics have been

shown to be inversely correlated with lignin content. Considering the quantitative importance of these by-products, considerable efforts have been made during the recent years to upgrade their nutritive value. Several pretreaments have been designed in order to improve the accessibility of cellulose and hemicelluloses to hydrolytic enzymes by disrupting the organization of the plant cell wall (Fan et al, 1982).

Lignin-degrading microorganisms and lignin-degrading enzymes have potential for selective delignification of lignocellulosic agricultural by-products. Various white-rot fungi have been studied for the purpose of selective delignification of straw as an alternative to chemical and physical pretreatments (Kirk and Moore, 1972; Zadražil 1980, Zadražil and Brunnert, 1981). Other studies have concerned the conversion of wood into feed (Reade and Mc Queen, 1983). Several strains of white-rot fungi remove lignin with limited degradation of cellulose and hemicelluloses (Abbott and Wicklow, 1984; Agosin and Odier, 1985b). Improvement of digestibility of wheat straw has been obtained using Dichomitus squalens (Agosin and Odier, 1985b, Zadražil and Brunnert, 1982) and Cyathus stercoreus (Wicklow et al, 1980) in semi-solid cultures. However the long cultivation time required to achieve this result casts doubt on its economic feasability unless a better control of delignification in semi-solid cultures is obtained.

RESULTS

The conversion of straw into a high quality feed using delignification by white-rot fungi in semi-solid cultures implies maximum lignin biodegradation with minimal cellulose and hemicellulose degradation. A selection of 75 strains of white-rot fungi has been conducted in order to evaluate the ligninolytic capacity of these strains as well as the degradation of cellulose and hemicelluloses (Agosin et al, 1985a). Biodegradation in wheat straw was estimated on the basis of $^{14}CO_2$ determination in cultures containing ^{14}C-(lignin) wheat straw. Total wheat straw degradation ability was assessed through determination of $^{14}CO_2$ evolved from ^{14}C-(whole labelled) wheat straw. Pycnoporus cinnabarinus, Pleurotus ostreatus, Dichomitus squalens Cyathus stercoreus Vavaria effuscata and Bjerkandera adusta were selected for their good capacity to degrade lignin and high selectivity for lignin removal (Agosin et al, 1985a). Phanerochaete chrysosporium (presiously named Sporotrichum pulverulentum) was the fastest strain for ligninolytic activity but cellulose and hemicellulose degradation by this organism was extensive.

Further study of wheat straw degradation and evolution of digestibility in semi-solid cultures with wheat straw showed that only Dichomitus squalens and Cyathus stercoreus were efficient in improving In Vitro Dry Matter Digestibility (IVDMD) with minimal dry matter loss (Figure 1) (Agosin and Odier, 1985b). Increase of IVDMD of wheat straw in cultures of Dichomitus squalens and Cyathus stercoreus were 6.25 %/day and 5.0%/day respectively. Increase of IVDMD took place during the early stages of decay after a lag time of 3-4 days. IVDMD of straw decayed by these fungi reached 63 and 68% repectively against 38% for the sound straw with dry matter loss of 15-20% (Agosin and Odier, 1985b).

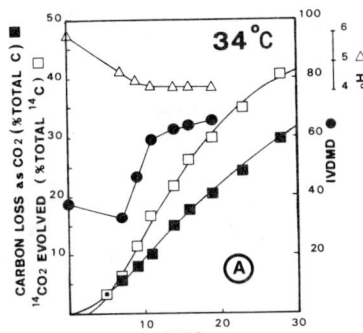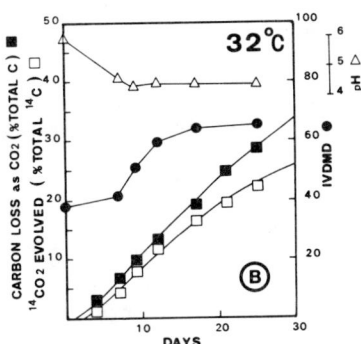

Figure 1: Carbon-loss as CO_2 (■), $^{14}CO_2$ from ^{14}C-lignin-wheat straw (□), IVDMD (●), and pH (Δ) in semi-solid cultures with wheat straw using Dichomitus squalens (A) and Cyathus stercoreus (B). Incubation took place at indicated temperature. From Agosin and Odier, 1985b.

In the case of Cyathus stercoreus increase of IVDMD was attained after 15 days incubation; at this stage the dry matter loss was 20% with 38% lignin degraded and 15.6 and 22.8% for cellulose and hemicellulose respectively.

Structural modifications in wheat straw during decay by the selected strains were analysed. All components of wheat straw were degraded in cultures. All studied strains degraded lignin in preference to cellulose and hemicelluloses. p-Coumaric and ferulic acid were degraded extensively in the early stages of decay.

Hemicelluloses were degraded in preference to cellulose by Cyathus stercoreus as well as Dichomitus squalens compared to cellulose as opposed to Phanerochaete chrysosporium which degraded cellulose in parallel to hemicelluloses. Uronic acids were only slightly degraded by the former strains while arabinose content decreased rapidly leading to a progressive enrichment of the remaining xylan in uronic acids with a parallel decrease of arabinose (Figure 2).

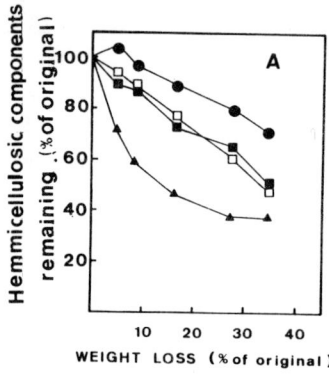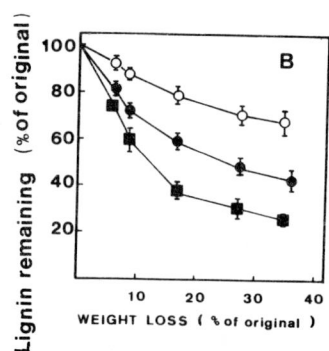

Figure 2: Degradation of hemicellulosic components (A) and lignin (B) during the semi-solid state fermentation of wheat straw using Cyathus stercoreus; A: arabinose (▲), O-acetyls (□), uronic acids (●) and xylose (■). B: Klason lignin (●), acid detergent lignin (■) and ^{14}C from lignin not converted to $^{14}CO_2$ (O). From Agosin et al, 1985c.

The removal of arabinose substituents in arabinoxylan may contribute to the increase of digestibility of xylan since it has been shown that arabinose removal precedes xylan digestion in the rumen (Dekker and Richards, 1976). In addition it has been shown that C-5 substituted arabinose which shows resistance to degradation in the rumen are bound to phenolic material by alkali-labile bonds while xylose units with this type of substitution are freely digestible (Chesson et al, 1983).

Analysis of methylated polysaccharides in straw decayed by Cyathus stercoreus and Dichomitus squalens showed a relative decrease of C-2 and C-3 in chain substituted xylose units as well as arabinose units. Whether C-5 substituted arabinose units are bound to lignin has not been firmly determined and whether the existence of a complex between arabinose and lignin restricts digestibility is still unknown.

Dichomitus squalens and Cyathus stercoreus degraded lignin with formation of high amounts of water-soluble compounds derived from lignin during the early stages of decay which were degraded upon further incubation. Gel permeation analysis showed that these compounds are largely polymeric (Figure 3). The formation of water-soluble polymeric material from lignin appears to be correlated with improvement of digestibility. Whether these compounds are digestible is unkown.

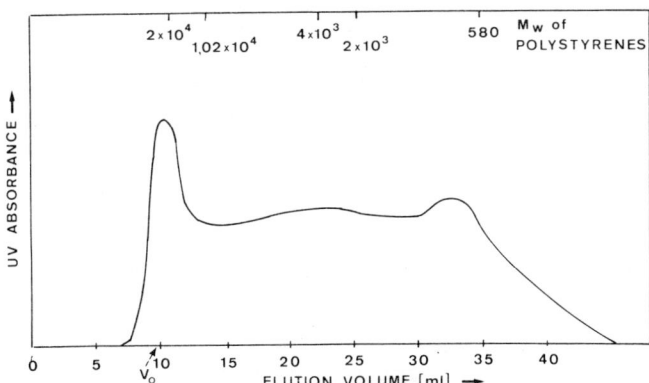

Figure 3: Elution profile in gel chromatography of water-soluble compounds formed from extracted wheat straw decayed by Cyathus stercoreus on 1.5 x 50cm Sephadex LH-60 with LiCl 0.1 M/ DMF solvent.

Water-soluble substances increased during decay by all white-rot fungi although variations were observed among studied strains. In cultures with Cyathus stercoreus water-soluble substances increased from 7% to 25%. Monomeric and oligomeric sugars represented 25 to 40% of the total water-soluble components which were formed in the early stages of decay. Among soluble sugars glucose, arabinose and xylose showed the greatest increase. Mycelium in decayed straw accounted only for a small amount of total dry matter, not exceeding 15%, and therefore is not expected to contribute significantly to the increase of IVDMD. Acetyl groups (Morris and Bacon, 1977) as well as phenolic acids (Hartley, 1972) have been reported to be partly responsible for the restricted degradation of forages in the rumen.

40

In this study no correlation was found between digestibility and decrease of these components during fungal decay. Inversely good correlation was found between Klason lignin content and IVDMD (r= 0.97); correlation with Acid Detergent Lignin was poor (r= 0.86) suggesting that lignin solubilized in sulphuric acid is important in restricting digestibility.

Digestion of fungal decayed wheat straw in the rumen was studied using the nylon bag method. Disappearance in the rumen of cell wall components in straw decayed by <u>Dichomitus</u> <u>squalens</u> and <u>Cyathus</u> <u>stercoreus</u> was compared with sound straw. The results are shown in table I.

Table I: Lag time (t_o, h), digestion rate constant (k, $\%.h^{-1}$) and potential of digestion (P.D., %) of insoluble components from wheat straw decayed by three white-rot fungi. Digestion kinetics parameters were defined from the model proposed by Ørskov and Mc Donald (1979):
$$P= A(1- \exp(-k(t-t_o)))$$

Fungus	parameter	% of original			
		total insolubles	Glucose	Xylose	Arabinose
Control	t_o	5.9 ± 1.1	4.2 ± 0.6	3.8 ± 1.5	3.6 ± 1.9
	k	2.9 ± 0.2	3.6 ± 0.4	3.8 ± 0.6	4.2 ± 0.7
	P.D.	48.2 ± 1.2	58.3 ± 2.8	56.1 ± 2.3	60.5 ± 8.3
Cyathus stercoreus	t_o	10.2 ± 1.4	12.8 ± 1.3	5.4 ± 1.9	7.8 ± 3.2
	k	4.4 ± 0.3	5.6 ± 0.5	4.0 ± 0.6	3.0 ± 1.1
	P.D.	59.8 ± 2.7	71.3 ± 4.1	62.1 ± 3.5	90.8 ± 12.0
Dichomitus squalens	t_o	3.7 ± 1.1	4.8 ± 1.1	0.2 ± 0.5	0.3 ± 2.9
	k	4.0 ± 0.3	4.0 ± 0.5	6.9 ± 0.8	4.2 ± 1.4
	P.D.	62.6 ± 2.1	71.3 ± 3.0	59.4 ± 4.2	72.8 ± 5.2

The rate and extent of digestion of total dry matter straw decayed by these fungi was significantly increased. Extent of cellulose digestion was enhanced in decayed straw while the extent of xylan digestion remained unchanged. Digestion rates were affected in various ways depending on the fungal strain: straw decayed by

<u>Cyathus</u> <u>stercoreus</u> showed increased cellulose digestion rates (50% increase) while xylan was degraded 1.8 times faster in straw decayed by <u>Dichomitus</u> <u>squalens</u>.

Methylation analysis showed that xylose units in branch points were more resistant to digestion in fungal treated straw. About 30 % of the lignin in straw was solubilized in the rumen; this lignin solubilization was faster for straw decayed by <u>Cyathus</u> <u>stercoreus</u>. p-Coumaric and ferulic acids were not digested. These results show that the enhanced degradability of cellulose and hemicelluloses after decay by the studied strains accounts for 45% of improvement of digestibility while the complement 55% correspond to formation of soluble substances assumed to be digestible.

CONCLUDING REMARKS

Although white-rot fungi are able to degrade all wheat straw components, certain strains such as <u>Dichomitus</u> <u>squalens</u> and <u>Cyathus</u> <u>stercoreus</u> degrade lignin more selectively with limited degradation of cellulose and xylan resulting in a better degradability of these components. Conversion of straw components into water-soluble compounds contribute significantly to improvement of feed value. Water-soluble compounds released during decay of wheat straw by white-rot fungi comprise soluble sugars arising from degradation of cellulose and xylan as well as lignin degradation products. The rapid removal of arabinose units in branch position in arabinoxylan during decay apparently improves further degradation in the rumen. It is proposed that improvement of feed value of wheat straw by white-rot fungi involves modification of xylan structure and not only lignin biodegradation.

ACKNOWLEDGEMENTS

This study has been financed by the Sanofi Santé Animale company.

REFERENCES

Abbott, P. A. and Wicklow D. T.1984. Degradation of lignin by
    Cyathus species. Appl. Env. Microbiol., 47, 585-587.

Agosin, E., Daudin, J.-J. and Odier, E. 1985a, Screening of white-
    rot fungi on (^{14}C) lignin -labelled and (^{14}C) whole-labelled
    wheat straw. Appl. Microbiol. Biotechnol., 22, 132-138.

Agosin, E., and Odier, E. 1985b. Solid-state fermentation, lignin
    degradation and resulting digestibility of wheat straw
    fermented by selected white-rot fungi, Appl. Microbiol. and
    Biotechnol., 21,397-403.

Agosin, E., Monties, B. and Odier, E. 1986. Structural changes in
    wheat straw components during decay by lignin-degrading white-
    rot fungi in relation to improvement of digestibility for
    ruminants, J. Sci. Food Agric., in press

Chesson, A., Gordon, A. H., Lowax, J. A. 1983, Substituents groups
    linked by alkali-labile bonds to arabinose and xylose residues
    of legume, grass and cereal straws and their fate during
    digestion by rumen microorganisms. J. Sci. Food Agric.,34,
    1330-1340.

Dekker, R.F.H. and Richards, G.N. 1976. Hemicellulases: their
    occurence, purification, properties and mode of action. Adv.
    Carbohydr. Chem., 32, 277-352.

Fan, L.T., Yong-Hyun Lee and Gharpuray, M.M. 1982. The nature of
    lignocellulosics and their pretreatments for enzymatic
    hydrolysis, Adv. Biochem. Eng., 23, 157-187.

Hartley, R. D. 1972. p-coumaric and ferulic acid components of cell
    walls of ryegrass and thier relationship with lignin and
    digestibility. J. Sci. Food Agr., 23, 1347-1354.

Kirk, T.K., and Moore, W.E. 1972. Removing lignin from wood with
    white-rot fungi and digestibility of resulting wood, Wood
    Fiber, 4,72.

Morris, E. and Bacon, J.S.D. 1977. The fate of acetyl groups and
    sugar components during the degradation of grass in sheep. J.
    Agric. Sci. Camb., 89, 327-340.

Ørskov, E.R. and Mc Donald 1979. I. The estimation of protein
    degradability in the rumen from incubation measurements
    weighed according to rate of passage. J. Agric. Sci. (Camb.),
    92, 499-503.

Reade, A.E., and McQueen, R.E. 1983. Investigation of white-rot
    fungi for the conversion of poplar into a potential feedstuff
    for ruminants. Can. J. Microbiol., 29,457-463.

43

Sundstol, F. and Owen, E. 1984. Straw and other fibrous by-products, Sundstrol and Owen, eds, Elsevier, Amsterdam.

Wicklow, D.T., Detroy, R.W. and Jesse, B.A. 1980. Decomposition of lignocellulose by Cyathus sterocreus (Schm) de Toni NRRL 6473, a "white rot" fungus from cattle dung. Appl. Environ. Microbiol., 40, 169-170.

Zadražil, F. 1980. Conversion of different plant waste into feed by basidiomycetes. Eur. J. Appl. Microbiol. Biotechnol.. 9,243-238.

Zadražil, F., and Brunnert, H. 1981. Investigation of physical parameters important for the solid-state fermentation of straw by white-rot fungi, Eur. J. Appl. MIcrobiol. Biotechnol., 11,183-188.

Zadražil, F., and Brunnert, H. 1982. Solid state fermentation of lignocellulose containing plant residues with Sporotrichum pulverulentum Nov and Dichomitus squalens (Karst) Reid. Eur. J. Appl. Microbiol. Biotechnol., 16, 45-51.

IMPROVEMENT OF OLIVE CAKE AND GRAPE BY-PRODUCTS FOR ANIMAL NUTRITION

J.F. Aguilera

Estación Experimental del Zaidín (CSIC)
18008-Granada (Spain)

I. Olive by-products

Annual world olive production amounts to 8.400.000 tonnes, which give rise to 3.000.000 tonnes of crude olive cake after industrial processing. Differences in both the oil extraction procedures and the further treatment of olive cake modify the relative proportions of the fruit components in the residue and therefore its chemical composition. In addition, the source and stage of madurity of the fruit contributes to the great variability in the analytical composition usually found in the olive by-products (table I). Fat and crude fibre fractions are reported to show the highest variability. The high lignin content of olive cake and the fact that most of its total nitrogen is linked to the lignocellulose fraction are the two main factors which appear to limit the digestive utilisation of olive residues.

Available data on digestibility of olive by-products are very scarce and extraordinarily heterogeneous (some by-products have been poorly described giving rise to contradictory results), but anyway they reflect a fairly low nutritive value (table II). It should be emphasized that care must be taken in considering particular results as optimal fermentation in rumen may have not probably been achieved in all cases.

The voluntary intake of olive cakes in sheep is rather high (85-140g DM per kg, $^{0.75}$) as reported in the literature (BOZA et al., 1970; ERASO et al., 1978; NEFZAOUI, 1985), although it is not yet enough to meet their maintenance requirements. This great consumption together with the small size of particles of these by-products (0.5-4 mm) results in a high rate of passage. In this context, NEFZAOUI (1985), using the Cr-mordanted fibre technique, obtained retention times of only 19-20 h, significantly lower than values over 40 h found for other forages (Van SOEST, 1982). Furthermore, the rate of degradation of olive cakes is slow and the values reported are fairly small. NEFZAOUI (1985) obtained, after 72 h of retention in rumen, 32% of degradation of dry matter working with a solvent-extracted olive cake.

The most efficient methods for improving the nutritive value of olive cakes are those that in a great extent decrease their high degree of lignification. Screening and alkali treatment seem suitable for this purpose (figure 1). Screening separates roughly the kernels or pits from the pulp. Alkali treatment dissolves hemicellulose and lignin, increasing both in vitro digestibility (MOLINA et al., 1984; NEFZAOUI, 1985) and in vivo digestibility (NEFZAOUI, 1985; AGUILERA and MOLINA, in press). NaOH-spray treatment followed by ensiling of a screened solvent-extracted olive cake (table III) brought about increases of up to 41 and 53% in digestible and metabolizable energy, respectively (AGUILERA and MOLINA, in press). In spite of the above mentioned processes the nutritive value of the treated products is lower than that of cereal straws, which questions the validity of these procedures.

The intake of screened and extracted olive cakes gave rise to low productions of VFA (50-65 mM/l), the VFA pattern being characteristic of forage resources; alkali treatment slightly increased VFA production (NEFZAOUI, 1985). Accordingly, CH_4 production attributable to the intake of a screened olive cake, measured in a respirometry chamber, was also found to be rather low (8.3-9.1 l/kg DM), tending to increase with the alkali treatment (AGUILERA and MOLINA, in press).

There is a limited research on the use of olive by-products as constituents of diets for ruminants of economic importance in which animal performance has been measured. Therefore, it is extremely difficult to draw conclusions from this evidence. The present recommendations of the Working Group for Improvement of Olive By-products (1983), sponsored by FAO, are summarised in table IV.

II. Grape by-products

Grape marc (winery pomace) is the main residue from wine making and consists of stalks (20-25%), grape pulp (45-50%) and grape seeds (15-25%). Annual world production of dried grape marc is estimated as 7.500.000 tonnes. Drying and ensiling are the suitable methods of preservation of these by-products.

A large variation in chemical composition has been found within and between grape by-products according to origin, technical procedure of juice extraction, further treatment of residues (factors which affect the pulp: seeds ratio) and preservation conditions (table V). Lignin content seems to vary to a large extent among by-products. The presence of both tannins (ranging from 3-7% DM) and links between protein and lignocellulose reduce dramatically the protein availability. N solubility is very low for extracted grape marcs, representing 1-2% total N (LARWENCE and YAHIAOUI, 1983; LARWENCE et al., 1983). This solubility is superior for grape pulp and surprisingly even higher for grape seed meals (10 and 37% total N, respectively), as reported by MORGAN and TRINDER (1980).

Table VI shows average results of digestibility of different by-products from wine making. The low organic matter digestibility of grape by-products can be attributed to both their high degree of lignification and their deficit of soluble N. Alkali treatment fairly improves the digestibility of grape marc (LARWENCE et al., 1983), this effect being not demonstrated in grape seed meal (COTTYN et al., 1981). As N content of grape by-products is not enough to cover rumen microflora requirements, the supplementation with a source of available N is expected to cause significant increases of the organic matter digestibility, as observed by LARWENCE and BERTHE (1981) and LARWENCE and YAHIAOUI (1983).

The experimental results of REYNE and GARAMBOIS (1977), LARWENCE et al. (1983), LARWENCE and YAHIAOUI (1983) and those reported by EZN (1981) show that in sheep ad libitum intakes of extracted grape marc silages are high (88-129 g per kg, $W^{0.75}$), but insufficient to meet maintenance needs.

There is very little information in the literature derived from the use of grape by-products in feeding trials. In dairy cattle the replacement of 20% concentrate by grape marc caused a decline in production, even when N content was adequately corrected (ECONOMIDES, 1974). On the contrary, in beef cattle the inclusion of either 15-30% grape (HADJIPANAYIOTOU and LOUCA, 1976) or 10% grape seed meal (COTTYN et al., 1981) led to satisfactory results. The use of grape by-products seems to be more attractive in sheep as higher levels of inclusion have proved adequate for moderate productions (SANCHEZ VIZCAINO and SMILG, 1971; REYNE and GARAMBOIS, 1977; LARWENCE et al., 1983). In this case, special attention should be paid to the Cu content of these residues.

Conclusions

Olive and grape by-products have many features in common: They are poor sources of energy and protein as indicated by their high degree of lignification, low digestibility and low content of available N. The use of proximate composition has proved inadequate to estimate their nutritional value. Althogh the rate of rumen degradation of these residues seems to be slow, their voluntary intake is rather high because of their small particle size. When fed to ruminants they require additional N to satisfy rumen microbial needs.

Olive and grape residues can be improved by alkali treatment providing a supply of degradable N is given. However, the energy content of the treated by-products appears to be lower than that of cereal straws, a fact which question the practical valitity of such a treatment. There is an urgent need to perform further experiments to measure any improvement in digestibility and animal production derived from chemical treatment and/or supplementation of these residues before solide conclusions can be drawn. Up to now the available data suggest that the use of olive and grape by-products should be restricted to minor components for low-output rations.

REFERENCES

AGUILERA, J.F. and MOLINA, E. (in press).

BOZA, J., FONOLLA, J. and AGUILERA, J., 1970. Rev. Nutr. Anim., 8, 13.

COTTYN, B.G., BOUQUE, Ch.V., AERTS, J.V. and BUYSSE, F.X., 1981. Agric. Environm., 6, 283.

ECONOMIDES, S., 1974. The effect of dried citrus pulp and grape marc on milk yield and milk composition of dairy cows. Tech. Paper Agric. Res. Inst. Nicosia, Cyprus, No. 7.

ERASO, E., OLIVARES, A., GOMEZ CABRERA, A., GARCIA DE SILES, J.L. and SANCHEZ, J., 1978. In: Nuevas fuentes de alimentos para la producción animal. (ed. A. Gómez Cabrera and J.L. García de Siles). ETSIA, Córdoba, Spain, p. 25.

E.Z.N., 1981. Bagaço de uva (valor nutritivo do producto ensilado). Estaçao Zootécnica Nacional. Folha Técnica do Departamento de Nutriçao e Alimentaçao Animal. Portugal.

HADJIPANAYIOTOU, M. and LOUCA, A., 1976. Anim. Prod., 23, 129.

LARWENCE, A. and BERTHE, J.L., 1981. Bull.Soc.Hist.Nat.Afr.Nord. 69,61.

LARWENCE, A., HAMMOUDA, F. and GAOUAS, Y., 1983. Ann. Zootech., 32, 371.

LARWENCE, A., HAMMOUDA, F. and SALAH, A., 1984. Ann. Zootech., 33, 533.

LARWENCE, A. and YAHIAOUI, A., 1983. Ann. Zootech., 32, 357.

MOLINA, E., AGUILERA, J.F. and BOZA,J., 1984. In: Valorisation des sous-produits de l'Olivier. Reunion du groupe de travail organisée par le Projet Regional D'amelioration de la Production Oleicole. FAO. Madrid, 1983, p. 117.

MORGAN, D.E. and TRINDER, H., 1980. In: By-products and wastes in animal feeding. (ed. E.R. Orskov). British Society of Animal Production. Occasional Publication. No. 3, p. 91.

NEFZAOUI, A., 1985. Valorisation des residus lignocellulosiques dans l'alimentation des ruminants par les traitements aux alcalis. Application aux grignons d'olive. Doctoral Thesis. Univ. Louvain-la-Neuve, Belgium.

REYNE, Y. and GARAMBOIS, X., 1977. Ann. Zootech., 26, 471.

SANCHEZ VIZCAINO, E. and SMILG, M., 1971. Rev. Nutr. Anim., 9, 153.

VAN SOEST, P.J., 1982. Nutritional ecology of the ruminant. (ed. P.J. Van Soest). O & Books, Inc., Oregon.

Table I. Chemical composition of some by-products from olive oil extraction (% DM)

	Crude cake	Extracted cake	Screened cake	Screened extracted cake (1)	Pulp
Ether extract	6-22	4-9	9-31	2-6	26-36
Crude fibre	34-57	35-43	16-36	22-35	16-25
Total N	0.6-1.6	1.2-1.6	1.1-1.9	1.4-2.2	1.3-2.2
Ash	2.5-10	4.5-10	3.7-7.5	6-8	5-9
N-ADF/Total N x 100				95-72	
Dry matter (%) 65-90		85-90	80-95	85-95	60-80

(1) NDF = 83-69; ADF = 64-54; ADL = 33-24; Hemicellulose = 9-13; Cellulose = 15-28

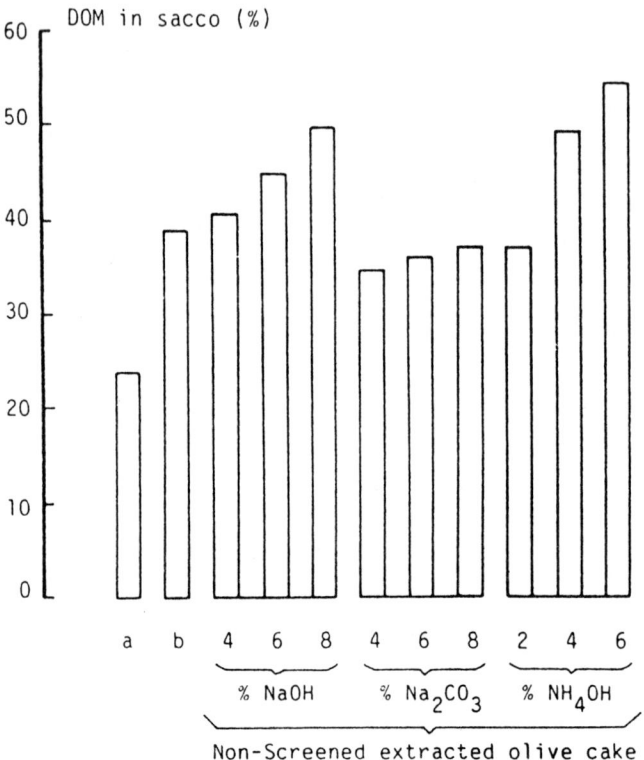

a = Non-Screened extracted olive cake
b = Screened extracted olive cake

Figure 1. Effect of screening or alkali treatment on the
         digestibility in situ of an extracted olive cake
         (Nefzaoui, 1985)

Table II. Coefficients of digestibility of some by-products from olive oil extraction obtained in sheep (%)

By-product	Approach	DM	OM	CP	EE	CF	ME*	References
Crude olive cake	Direct		30.8	6.6	65.5	28.4		(1)
	By difference	32.9	35.4	24.5	57.7	29.6	5.2	(2)
	By difference		32.5	10.0	86.8		4.8	(3)
	By difference		32.2	9.0	85.1	44.2	4.8	(3)
	By difference		45.7	23.6	75.2		6.7	(4)
Screened olive cake	Direct		37.2	19.4	84.1	33.6		(5)
	By difference		21.6	15.5	85.6	12.8	3.1	(6)
	By regression	41.9	49.9	32.5	91.5	22.2		(7)
Screened extracted olive cake	Direct	48.1	50.0	32.2	80.2	47.3		(8)
	Direct	30.5	32.2	38.8	81.8	22.5	4.7	(8)
	Direct	36.5	39.7	29.0	77.4	39.1	5.7	(8)
	Direct		48.0	52.1	77.8	47.9	6.9	(9)
	Direct		18.8	8.0	27.6	16.6	2.7	(9)
	By difference	36.1	36.7	15.8	74.1	2.1	5.5	(10)
	By difference	19.1		25.4	88.9	27.0		(11)
	By difference	26.3	26.9	4.3			4.2	(12)
Olive pulp	By difference	48.3		21.6	86.6			(13)
	By difference		43.7	13.4			6.4	(4)
	By difference		57.4	66.8	90.0		8.4	(4)
	By difference	50.5	51.9	9.9	88.0	57.0	7.4	(14)
	By difference	57.4	57.6	11.0	90.5	66.4	8.2	(14)
Extracted olive pulp	By difference		69.4	28.0			10.2	(4)

* Calculated according to ME (MJ/kg DM) = kg DOM x 19 x 0.82

(1) Kellner (1924); (2) Boza and Varela (1960); (3) Boza et al. (1970); (4) Theriez and Boule (1970); (5) Maymone and Carusi (1935); (6) Maymone et al. (1961); (7) Ben Hamouda (1975); (8) Nefzaoui (1985); (9) Eraso et al. (1978); (10) Accardi et al. (1979); (11) Valamotis (1983); (12) Aguilera and Molina (in press); (13) Maymone et al. (1962); (14) Duranti et al. (1978).

Table III. Digestibility and energy content of a screened solvent-extracted olive cake. Effect of NaOH treatment. (Aguilera and Molina, in press)

g NaOH/100g by-product	0	5	7.5	10	SEM	Level of significance
Dry matter	26.0^a	38.8^b	38.7^b	42.2^b	1.22	***
Organic matter	26.9^a	35.6^b	34.2^b	36.3^b	1.51	**
Crude protein	4.3^a	23.8^b	16.0^b	14.8^b	3.10	**
DE/GE	25.5^a	32.8^b	34.6^b	39.2^c	1.34	***
DE (MJ/kg DM)	5.41^a	6.66^b	6.98^b	7.61^c	0.27	***
ME/GE	19.9^a	28.3	30.1^b	33.3^c	1.61	***
ME (MJ/kg DM)	4.22^a	5.75^b	6.07^b	6.46^c	0.32	***
ME/DE	77.7^a	86.5^b	86.9^b	84.7^c	2.39	***

** = P<0.01; *** = P<0.001

Values in the same row bearing different superscripts differ significantly (P<0.05)

Table IV. Recommendations for the use of by-products from olive oil extraction in practical diets for ruminants

By-product	Level of production			
	Survival	Maintenance	Moderate	Intensive
Extracted olive cake	Ad lib + forage +.....	NR[1]	NR	NR
Crude olive cake	Ad lib + forage +.....	Ad lib + forage +..... <30%	NR	NR
Screened olive cake	Ad lib + forage +.....	Ad lib + forage +..... <30%	<30%	NR
Screened extracted olive cake	Ad lib + forage +.....	Ad lib + forage +..... <30%	40-50%	NR
Olive pulp	Ad lib + forage +.....	<30%	<30%	NR

(1) NR = Not recommended

Table V. Chemical composition of some by-products from wine making (% DM)

By-product	DM(%)	CP	EE	CF	Ash	NDF	ADF	MADF	ADL	References
Grape marc	46.5	11.5	12.2	26.0	5.1					(1)
		13.7	7.0	23.6	12.8					(2)
Grape marc silage	31.4	12.9	7.6	32.1	6.9					(3)
Dried grape marc	92.5	12.3	8.5	35.4	4.6					(4)
	91.0	13.4	7.5	33.2	5.5					(5)
Dried steam extrated grape marc	87.2	12.0	4.7	20.4	5.6					(6)
	90.5	11.8	11.0	26.3	5.3					(7)
Water extracted grape marc	32.7	11.9		29.0	5.6					(8)
Steam extracted grape marc	32.8	13.1	8.3	24.2	6.8					(8)
Steam extracted grape marc silage	43.1	14.2	8.3	28.9	5.3	74.2	59.4		40.3	(9)
	44.7	14.4	7.8	27.0						(10)
Not extracted grape seeds	93.0	9.6	15.2	45.7	4.2					(2)
Grape seed oil meal, solvent	86.8	9.3	3.1	39.2	5.6			58.4		(11)
	85.5	11.0	0.7	53.3	3.2	81.1	67.0		13.7	(12)
		11.0	0.8	49.6	3.3					(12)
Grape seed oil meal, hydraulic	91.0	13.6	8.5	44.0	5.2					(2)
Grape pulp pellets		15.0	4.9	21.9	7.2			55.5		(11)
		15.1	6.4	22.7	4.7			56.4		(11)

(1) Mamaev et al. (1975); (2) Bo Gohl (1975); (3) EZN (1981); (4) Economides and Hadjidemetriou (1974); (5) Morrison (1957); (6) Dumont and Tisserand (1978); (7) Sánchez Vizcaíno and Smilg (1971); (8) Reyne and Garambois (1977); (9) Larwence and Yahiaoui (1983); (10) Larwence et al. (1983); (11) Morgan and Trinder (1980); (12) Cottyn et al. (1981).

Table VI. Coefficients of digestibility of some by-products from wine making obtained in sheep (%)

By-product	DM	OM	CP	EE	CF	NDF	ADF	ADL	ME*	References
Dried grape marc	32.2	29.6	0	71.2	16.6				4.8	(1)
	28.4	32.2	21.7						4.4	(2)
		29.7	19.0	91.0	10.1				6.7	(3)
		45.7	12.9	56	15					(4)
			12							(5)
Grape marc	22.9									(6)
Grape marc silage		22.0	9	45	35				3.2	(7)
Ensiled steam extrac-		28	13	48	26				4.0	(8)
ted grape marc (ESEGM)		25	8	62	16				3.6	(8)
		28	8		23				4.1	(8)
	31.5	32.2	9.7	46.0	31.7	27.5	29.0	26.3	4.8	
ESEGM + 0.67% NaOH	45.4	44.2	28.8	70.0	51.5	44.5	42.2	22.7	6.4	(9)
ESEGM + 1.11% NaOH	42.2	41.5	21.4	68.3	45.3	34.6	38.2	14.7	6.0	
ESEGM	28.2	30.8	19.3	48.9	27.5				4.1	
ESEGM + N source	40.3-50.1	40.6-50.6	37.9-48.5	55.1-68.1	33.8-51.0				5.9-7.4	(10)
ESEGM	29.8	31.1	10.1	46.1	27.4				4.6	(11)
ESEGM + 20g PEG 4000	37.4	38.7	28.8	48.2	30.3				5.7	
Grape seed oil meal (GSOM)	28.6	28.2	58.8	87.3	11.1	19.5			4.3	(12)
	24.7	24.9	56.6	100	25.1				3.8	
GSOM + 4% NaOH	26.2	22.5	48.7	30.3	20.5	22.2			3.2	(12)
Grape seed oil meal			34.0		5.5					(13)

* Calculated according to ME (MJ/kg DM) = kg DOM x 19 x 0.82

(1) Sarti (1970); (2) Sánchez Vizcaíno and Smilg (1971); (3) Economides and Hadjidemetriou (1974); (4) Dumont and Tisserand (1978); (5) Morrison (1957); (6) Bo Gohl (1975); (7) EZN (1981); (8) Larwence et al.(1983); (9) Reyne and Garambois (1977); (10) Larwence and Yahiaoui (1983); (11) Larwence et al. (1984); (12) Cottyn et al. (1981); (13) Ferrando and Catsaounis (1966)

White rot fungi and mushrooms grown on cereal straw: aim of the
process, final products, scope for the future.

F. Zadrazil

Institut für Bodenbiologie der Bundesforschungsanstalt für
Landwirtschaft, Bundesallee 50, D-3300 Braunschweig, FRG.

Summary

Different edible species of Basidiomycetes or Ascomycetes were
tested for mycelium growth and fruit body production and the increase in in vitro digestibility of wheat straw and other lignocellulosics (sunflower, reed, sawdust etc.). Fungal growth and
substrate changes were influenced by the nutrients in plant waste
material and cultivation conditions. The best edible fungi, which
give good yield of fruiting bodies and show an increase in in
vitro digestibility are: Pleurotus spp., Stropharia rugosuannu-
lata. Stropharia rugosuannulta increase digestibility of straw
from 40 to 72% and Pleurotus spp. to 60-65%.

Results and discussion

The main barrier of the biological decomposition of lignocellulosics is lignin and its complex structure with cellulose. Only
some microorganisms, mainly white rot fungi, are able to decompose all plant polymers simultaneously. Bacteria, yeasts and
moulds preferably metabolize monomers and require a chemical or
physical pretreatment of the lignocellulosic material for suc-
cessful growth.

Fig. 1 shows principal possibilities for the upgrading of plant
wastes into food and feed. It must be distinguished between feed
for ruminants, which can contain cellulose and hemicellulose and
feed for monogastrics, which mainly contains microbial proteins
and sugars. The use of lignocellulosic, only without pretreatment will be discussed later on.

Fig. 1

Different ways of microbial using of fibrous by-products
(lignocellulosic) for feed / food production

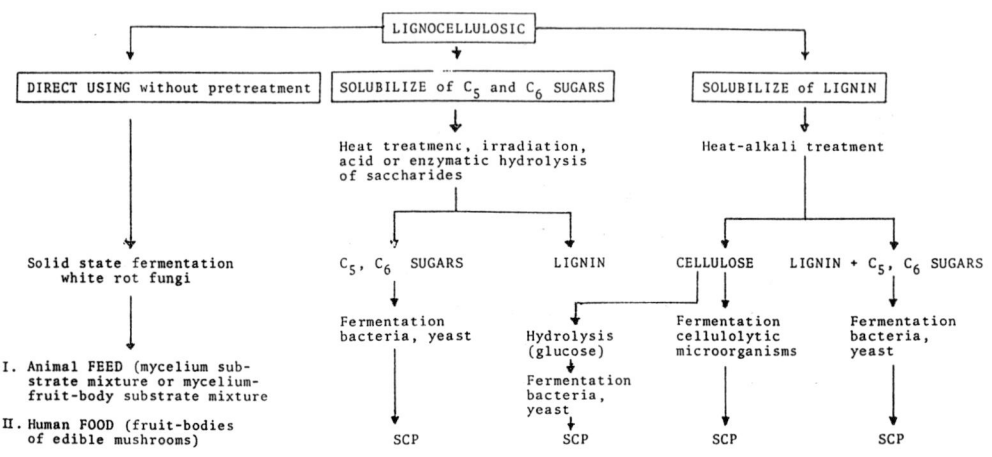

Conversion of lignocellulosics into edible mushroom

A great number of edible fungi can be cultivated on cereal straw.
Some growth properties and characteristics of 24 species are
numerically summarized in Table 1. The ability to grow and form
fruiting bodies depend on kind, composition and treatment of
lignocellulosics. (For a part of these fungi some additives into
the substrate are necessary to obtain good yields).

Conversion of lignocellulosics into feed

Higher fungi differ greatly in their ability to alter the digestibility of lignocellulosics for ruminants. Some of them show positive effects, whereas other have colonizing lignocellulosics or
do even show an adverse effect on the in vitro digestibility of
this substrate. Species Stropharia rugosoannulata, Pleurotus spp.
and Abortiporus biennis almost double the in vitro digestibility
of straw.

Table 1. Edible fungi which can be cultivated on cereal straw substrate (growth characteristics and digestibility of substrate).

Agaricus arvensis	1	5	6	9	12	15	17-18	19b	22	25
Agaricus bisporus	1	5	6	9	12	15	17	21a	23	25
Agaricus bitorquis	1	5	6	9	12	15	18!	21a	23	25
Agrocybe aegerita	1(2)	3	8	10	13	15	18!	20b	22	25
Auricularia judea	1	3	7	10	13	15	17-18	19b	22	25
Coprinus comatus	1	5	7	10	12-13	15	17	19a	23	25
Coprinus fimentarus	1	5	7	10	12-13	15	?	19a	?	?
Flammulina velutipes	2(1)	3	8	10	12-13	15	17	19b	22	25
Kuchneromyces mutabilis	1	3	8	10	12-13	15	17	19b	23	25
Lentinus edodes	1(2)	3	7	9	12-13	15	18	19-20b	23	26
Lepista nuda	1	5	6	10	12	15	17	19b	?	?
Macrolepiota procera	1	5	6	10	12	15	17	19b	23	?
Macrolepiota rhacodes	1	5	6	10	12	15	17	19b	?	?
Pleurotus sp. abalone	1(2?)	3	8	11	14	16	18!	20-21a	24	26
Pleurotus cornucopiae	1(2?)	3	8	11	14	16	18!	21a	24	26
Pleurotus eryngii	2(1)	4	8	10	13	16	18!	21a	24	26
Pleurotus sp. "Florida"	1(2?)	3	8	11	14	16	18!	21a	24	26
Pleurotus flabellatus	1(2?)	?	8	11	14	16	18!	20a	24	26
Pleurotus ostreatus	1(2?)	3	8	11	14	16	17	21a	24	26
Pleurotus sapidus	1(2?)	3	8	11	14	16	18!	21a	24	26
Pleurotus sajor caju	1(2?)	?	8	11	14	15	18!	20-21a	24	26
Pholiota nameko	1	3	7	10	13	15	18	19-20b	22	25
Stropharia rugosoannulata	1	5	8	11	13	16	17	19-20a	24	26
Volvaria volvacea	1	5	8	11	13	15	18!	19-20a	22	25

Explanation to Table 1.

1 saprophyte
2 parasite

Substrate in nature

3 wood
4 perennial plants
5 soil org. matter or plant waste

Growth on sterile straw substrate
  (25% straw, 75% water)
6 slow

7 good
8 very good

Saprophytic colonization ability on
  pasteurized straw substrate

9 low
10 good
11 very good

Decomposition rate of straw substrate

12 low
13 medium
14 high

Formation of fruiting bodies and yield

15 for good yield additives into substrate are necessary
16 for good yield additives into substrate are not necessary

17 formation of fruiting bodies below 18°C
18 formation of fruiting bodies above 18°C
18! useful in subtropical and tropical countries
19 yield of fruiting bodies is relatively low
20 good
21 very good
a commercially cultivated on straw
b experimental stage

Decomposition of cereal straw-lignin

22 no or low
23 good
24 very good

Increase of digestibility of cereal straw

25 no or low
26 good or very good

Ganoderma applanatum increase the in vitro digestibility of hardwood "Palo podrido" by 77 % (Zadrazil et al. 1982). For more information about the upgrading of cereal straw by white rot fungi
have a lock at the following list of literature: Kirk and Moore
(19172), Hartley et al. (1974), Zadrazil (1977, 1979, 1980),
Kaneshiro (1977), Latham (1979), Lindenfelser et al. (1979),
Langar et al. (1980), Streeter et al. (1982), Ibrahim and Pearce
(1980). Cultivation on maple or poplar wood was performed by
Matteau and Bone (1980), McQueen et al. (1980) and on poultry
droppings by Virk et al. (1980).

Table 2 show the results of the in vitro digestibilities of different substrates after treatment with white rot fungi.

Table 2: Changes in digestibility of lignocellulosics by various fungi

Fungus	Temperature °C	Increase in digestibility from (%) to		Time of fermentation (days)	Substrate	Reference
1. Polyporus berkeleyi	27	46	77	88	Aspen wood	Kirk and
2. Polyporus frondosus	27	46	71	64	Aspen wood	Moore (1972)
3. Fomes ulmarius	27	46	64	77	Aspen wood	
4. Fomes lividus	32	5	26	14-21	Beech wood	
	32	14	36	14-21	Oak sap wood	Hartley et al.
5. Phellinus melanoporus	25	5	26	14-21	Beech wood	(1974)
	25	14	30	14-21	Oak sap wood	
6. Stropharia rugoso-annulata	22	40	55	79	Wheat straw	
	25	40	61	79	Wheat straw	Zadrazil (1977)
	30	40	69	79	Wheat straw	
7. Peniophora gigantea	14-25	36	46	21	Barley straw	
	14-25	32	39	21	Sugarcane bagasse	Ibrahim and Pearce (1980)
8. Ganoderma lucidum	14-25	40	48	21	Pea straw	
9. Peniophora cremea	14-25	20	27	21	Sunflower hulls	
10. Polyporus anceps	25	30	72	56	Poplar shaving	
11. Ganoderma applanatum	25	30	64	28	Poplar shaving	
12. Phanerochaete chrysosporium	25	30	62	28	Poplar shaving	Reade and McQueen (1983)
13. Polyporus versicola	25	30	61	21	Poplar shaving	
14. Fomitopsis ulmarius	25	30	42	28	Poplar shaving	
15. Sporotrichum pulverulentum	35	16	34	21	Wheat straw	Parven et al. (1983)

The results of the above mentioned publications can be summarized as follows:

1. The in vitro digestibility of fungal substrates decreases at the beginning of the cultivation of lignocellulosics by white rot fungi and increases afterwards (Zadrazil 1977, and Zadrazil and Brunnert 1982).

2. The increase in in vitro digestibility depends on the fungal species (Zadrazil 1979, 1983), cultivation time and temperature, the water/air ratio and gas composition (O_2; CO_2; N_2) in the substrate, and on the preparation, bulk density and composition of the substrate (Zadrazil and Brunnert 1981, 1982).

3. The in vitro digestibility of lignocellulosics by white rot fungi is decreased by inorganic nitrogen additives (Zadrazil and Brunnert, 1980).

4. During the incubation of lignocellulosics the content of soluble substances (partly sugars) increases (Zadrazil 1976, Lindenfelser et al. 1979).

The above mentioned experiment have mostly been carried out on a laboratory scale. However, little is known about the possibilities of transforming these results to large scale.

The analyses of "palo podrido" (palo podrido white coloured, decomposed wood, used as animal feed) clearly show that at least in the laboratory scale and under natural conditions the use of white rot fungi for the upgrading of lignocellulosics into feed, is possible (Zadrazil et al. 1982).

On the other hand only little is known about scale up experiments and the large scale process. One such possibility could be the fermentation, as already applied for the production of substrates for Agaricus spp. and Pleurotus spp..

Conclusion

Solid state fermentation of lignocellulosics with white rot fungi
is a polyfactorial process with a limited number of control-poss-
ibilities. Fungal species, time and temperatur of fermentation,
the kind and chemical composition of substrate and its physical
structure as well as the composition of the gas phase (O_2; CO_2;
N_2) within the substrate are the most important factors con-
trolling lignin degradation and in vitro digestibility of the
fermented product.

For practical applications research concerning the following
fields is needed:

  a) Development of solid state fermentation process

  b) Scale up experiments by constant product quality and high
     yield of fruiting bodies as in laboratory scale

  c) Quick methods for the determination of the quality of sub-
     strate mycelium mixture (in vitro digestibility) and for
     conrolling the fermentation process

More details of this project will be discussed on the proposed
workshop in Septempber 1986 in Braunschweig.

References

Hartley, R. D.; Jones, E. C.; King, N. J. and Smith, G. A., 1974.
Modified wood waste and straw as potential components of animal
feed. J. Sci. Fd. Agric., 25: 433-437.

Ibrahim, M. N. M. and Perce, G. R., 1980. Effects of white rot
fungi on the composition and in vitro digestibility of crop by-
products. Agricultural wastes., 2: 199-205.

Kaneshiro, T., 1977. Lignocellulosic agricultural waste degraded
by Pleurotus ostreatus. Dev. Ind. Microbiol., 18: 591-597.

Kirk, T. K.; Moore, W. E., 1972. Removing lignin from wood with
white-rot fungi and digestibility of resulting wood. Wood Fiber.,
4: 72-79.

Langar, P. N.; Seghal, J. P. and Garcha, H. S., 1980. Chemical
changes in wheat and paddy straw after fungal cultivation. Indian
J. Anim. Sci., 50: 942-946.

Lindenfelser, L. A.; Detroy, R. W.; Ramstack, J. M. and Worden, K. A., 1979. Biological modification of the lignin and cellulose components of wheat straw by Pleurotus ostreatus. Dev. Ind. Microbiol., 20: 541-551

Matteau, P. P.; Bone, D. H., 1980. Solid-state fermentation of maple wood. Biotechnol. Letters., 2: 127-132

McQueen, R. E and Reade, A. E.,1980. Changes in composition and digestibility of poplar by fungal fermentation. Can. J. Animal Sci., 60: 571-572.

Reid, I. D., 1979. The influence of nutrient balances on lignin degradation by the white-rot fungus Phanerochaete chrysosporium. Can. J. Bot., 57: 2050-2058.

Rypacek, V., 1966. Biologie holzzerstörender Pilze. VEB, G. Fischer, Jena.

Streeter, C.L.; Conway, K.E.; Horn, G. W. and Mader, T. L., 1982. Nutritional evaluation of wheat straw incubated with the edible mushroom, Pleurotus ostreaus. J. anim. Sci., 54: 183-188.

Virk, R. S.; Sethi, R. P.; Garcha, H. S., 1980. Note on the conversion of poultry droppings by Pleurotus ostreatus into feed. Indian J. Anim. Sci., 50: 293-295.

Zadrazil, F., 1976 b. Freisetzung wasserlöslicher Verbindungen während der Strohzersetzung durch Basidiomyceten als Grundlage für eine biologische Strohaufwertung. Z. f. Acker- und Pflanzenbau., 142: 44-53.

Zadrazil, F., 1977. The conversion of straw into feed by Basidiomycetes. Eur. J. Appl. Microbiol., 4: 291-294.

Zadrazil, F., 1979. Umwandlung von Pflanzenabfall in Tierfutter durch Pilze. Mushroom Sci. (Part I.), X: 231-241.

Zadrazil, F., 1980. Conversion of different plant wastes into feed by basidiomycetes. Eur. J. Appl. Microbiol. Biotechnol., 9: 243-248.

Zadrazil, F., 1983. Screening of fungi for lignin-decomposition and conversion of straw into feed. Eur. J. Appl. Micorbiol. Biotechnol., (in press).

Zadrazil, F. and Brunnert, H., 1980. The influence of ammonium nitrate supplementation on degradation and in vitro digestibility of straw colonized by higher fungi. Eur. J. Appl. Microbiol. Biotechnol., 9: 37-44.

Zadrazil, F. and Brunnert, H., 1981. Investigation of physical parameters important for the solid state fermentation of straw by white rot fungi. Eur. J. Appl. Microbiol. Biotechnol., 11: 183-188.

Zadrazil, F. and Brunnert, H., 1982. Solid state fermentation of lignocellulose containing plant residues with Sporotrichum pulverulentum Nov. and Dichomitus squalens (Karst.) Reid. Eur. J. Appl. Microbiol. Biotechnol., 16: 45-51.

Zadrazil, F.; Grinbergs, J. and Gonzalez, A., 1982. "Palo podrido" - decomposed wood which was used as feed. Eur. J. Appl. Microbiol. Biotechnol., 15: 167-171.

Zadrazil, F., 1985. Screening of fungi for lignin-decomposition and conversion of straw into feed. Angewandte Botanik, Heft 5/6.

COMPOSITION OF THE NON-STARCHY POLYSACCHARIDES
OF PROTEIN EXTRACTION RESIDUES FROM LEGUME SEEDS

J. DELORT-LAVAL, M. CHAMP

NATIONAL INSTITUTE OF AGRICULTURAL RESEARCH

Laboratory of Animal Feed Technology

Rue de la Géraudière - 44072 NANTES CEDEX - FRANCE

ABSTRACT

Protein concentration from legume seeds or oilseeds usually implies pre-liminary steps, such as dehulling, oil and soluble carbohydrates extraction. The main coproducts of protein separation are the seed hulls and solid resi-dues from protein extraction (wet process) or from grinding and air classifi-cation (dry process); residues from starchy material contain mainly starch and a small amount of cell wall polysaccarides. Recent results on the detailed composition of legume seeds hulls and cotyledon cell wall non starchy polysac-charides are presented.

INTRODUCTION

Some of the proteins of plant origin which could be used directly for hu-man consumption have an aminoacid composition which can be compared to that of animal protein sources. However, plant proteins are not greatly appreciated by the population of industrialized countries because of the protein dilution by large amounts of various carbohydrates.

Protein extraction processes have mainly been developed from legume seeds (GUEGUEN, 1983; COLONNA et al, 1980), oilseed meals (sunflower and rapeseed (FRANCHINO et al, 1983; FANTOZZI and SENSIDORI, 1983) and leaves (FIORENTINI and GALOPPINI, 1983). They are described in a general review on the extraction and purification technologies of vegetal protein matters (BEROT and DAVIN, 1985). In the present paper, description of protein extraction process will be limited to legume seeds.

PROTEIN EXTRACTION PROCESSES FROM LEGUME SEEDS

Legume seeds are characterized by their high protein level. Some of them (groundnut, soybean) are mainly used in animal feeding as oilseed cakes or meals. These meals can constitute the basic raw material for preparation of

protein concentrates or isolates used in human food products. Several processes of protein extraction have also been developed from starchy legume
seeds (mainly fababean and pea).

The processes generally used are (1) pin-milling plus air-classification
which when applied to starch-rich legume seeds, results in concentrates containing 60-75% proteins or (2) wet processes which produce isolates (90-95%
proteins) (GUEGUEN, 1983). The starchy legume seeds have to be selected, dehulled and ground before protein extraction.

Dry process is not efficient for soybean and lupine flours because of
their low content in starch (SOSULSKI and YOUNGS, 1979). SOSULSKI and YOUNGS
(1979) have studied the efficiency of pin-milling and air classification on
eight starchy legume seeds (chick pea, pea, northern bean, fababean, field
pea, lima bean, mung bean and lentil). The light material which represents
22.5% to 29% of the original flours has a protein content of 47,7% to 66,6%
(except for chick pea) and contains most of the flour lipids and ash while the
coarse fraction contains 51% to 68% starch and cell wall material and only
12,2% to 15,6% protein.

In order to prepare protein isolates by wet process, the most widely used
method is the one patented by ANSON and PADER (1957). After an alkaline solubilization of the proteins, the insoluble material is removed by centrifugation. By adding hydrochloric acid to the supernatant, the protein is precipitated isoelectrically, thereby producing the isolate. From dehulled fababean,
pea and other seeds with a high starch content, the byproducts of wet protein
extraction are mainly constituted of starch and pectic substances (BRILLOUET,
1982). When this process is applied to lupine or soybean defatted flour, the
main byproduct contains a large amount of fibers.

COMPOSITION OF LEGUME SEEDS HULLS (Table I)

For most legume seeds, hulls are a minor part of the seed (Table I). They
are generally obtained by dehulling whole or coarsely ground seeds. They are
largely indigestible by monogastric animals (BAILEY et al, 1974), but could be
used in ruminant feeding. They are poor in lignin (1.2 to 2.1%) and very rich
in cellulose (41 to 29% cellulosic glucose).

SANNELLA and WHISTLER (1962) have isolated from soybean hulls a polysaccharide (hemicellulose B) by delignification, potassium hydroxyde extraction and 70% ethanol precipitation of the supernatant. This polysaccharide (yield : 6%) contains xylose, arabinose, galactose, glucose and glucuronic acid in molar ratios 14:1:3:3:3. Most of the xylose units are engaged in the basic backbone which is branched with the other sugars. Hemicellulose A (yield : 3.6%) fraction, obtained by potassium hydroxyde extraction and copper-complexing of an oxalate-EDTA extracted soyabean hull, is a glucuronoxylan devoid of arabinose (ASPINALL et al, 1966).

Total hemicelluloses constitute 13% of the sweet lupine hulls. Hemicellulose A (70% of total hemicelluloses) contains mainly xylose, whereas arabinose (mainly), glucose, galactose and xylose (traces) are detected in hemicellulose B hydrolysates (BAILEY et al, 1974). A part of hemicelluloses lupine hulls is mainly composed of 51% cellulose, 28% pectic substances (18% oxalate-soluble polyuronide and 10% oxalate-soluble neutral hexose polymers) (BAILEY et al, 1974). The composition in neutral and acidic sugars of the hull of three starchy legume seeds has been compared by CHAMP et al (1986) (Table I). Uronic acids, which are supposed to be of both glucuronic and galacturonic type constitute from 11% (lentil) to 25% (chick pea) of the hulls. Non cellulosic neutral sugars are mainly arabinose and xylose. No structural information was reported for these legume seed hulls.

COMPOSITION OF COTYLEDON CELL WALL POLYSACCHARIDES (Table II)

Byproducts of wet protein extraction from dehulled seeds contain the main part of the cell wall polysaccharides of the kernel, whereas the dry process does not concentrate these polysaccharides in the starchy fraction. COLONNA et al (1980) have compared the composition of the starchy byproduct obtained after dry and wet protein extraction from smooth or wrinkled pea and fababean. The first fractionation procedure (dry process) of the seeds includes fine milling and air classification. The dense material contains most of the starch, but only a small fraction of the cellulosic and hemicellulosic material. With broad beans, cellulose, lignin and pentosans decrease from 1.0,0.7 and 2.0% in flour to 0.26%, 0.04 and 0.35% respectively in the starchy fraction. With smooth pea, the cellulose content increases from 0.9% in flour to

1.7%, whereas the lignin and pentosans contents decrease from 0.5 and 2.3 to 0.2 and 0.6% respectively. In the case of wrinkled pea, cellulose, pentosan and lignin decrease from 1.2-3.3-0.3 to 0.3-3.0-0.1 respectively.

In the wet process, starch is again the main residue, but non starchy polysaccharides are more concentrated than in the starting material. Based upon recoveries of the starchy byproducts, smooth pea and fababean lignin is fully recovered; furfural generators are partly (15%) lost during smooth pea processing; cellulose is only partly recovered in smooth pea (67%) and broad beans (53%). For broad beans, the molar ratios of polysaccharides constituent sugars, relative to xylose, remain constant for arabinose, rhamnose and ribose, but decrease in the byproduct for mannose and galactose. For smooth pea, the ratios vary : arabinose/xylose and rhamnose/xylose increase, whereas other ratios decrease (COLONNA et al, 1980).

A wet sieving process of protein extraction has been described for dehulled lupines (DAVIN and BRILLOUET, 1984), yielding 12,5 (in L.mutabilis) to 28% fiber (in L.albus) from the initial meals. They mainly contained pectic polysaccharides (BRILLOUET and RIOCHET,1983), consisting predominantly of galactose, arabinose and galacturonic acid residues (Table II). Similar analysis has been made on kidney bean, lentil and chickpea cell walls (CHAMP et al, 1986). Arabinose, glucose and galacturonic acid are the main sugars of the three seed polysaccharides.. However, kidney bean cell walls contain a higher xylose content (6.9%) than those of lentil (3.4%) or chick pea (2.4%). Most of the cell wall glucose comes from betaglucans.

CONCLUSIONS

Byproducts from protein extraction from legume seeds are mainly hulls and starchy or non starchy polysaccharides residues. Hulls are rich in cellulose and slightly lignified; they can be used efficiently in ruminant feeding. Starchy byproducts obtained from protein extraction of dehulled legume seeds contain only limited amounts of cell wall material; they are potential sources of digestible carbohydrates in the one-stomach animal. Byproducts of non starchy seeds, such as soybean or lupine, contain mainly pectic polysaccharides, which are highly fermentescible. They even could be implied, with the alpha-galactosides, in the flatus observed in pigs fed diets rich in lupine meal.

REFERENCES

ANSON M.L., PADER M., 1957 - Extraction of soy protein, US Pat. 2 785 155.

ASPINALL G.O., HUNT K., MORRISON I.M., 1966 - Polysaccharides of soybeans. Part II - Fractionation of hull cell wall polysaccharides and the structure of a xylan. J. Chem. Soc., C, 1945-1949.

BAILEY R.W., MILLS S.E., HOVE E.L., 1974 - Composition of sweet and bitter lupine seed hulls with observations on the apparent digestibility of sweet lupine seed hulls by young rats. J. Sci. Food Agric., 25, 955-961.

BEROT S., DAVIN A., 1985. Technologie d'extraction et de purification des matières protéiques végétales. In "Les protéines végétales : aspects biochimiques, technologiques, nutritionnels et économiques", B. GODON éd., Lavoisier, Paris.

BRILLOUET J.M., 1982. Non starchy polysaccharides of legume seeds from the papilionoideae sub-family. Sci. Aliments, 2 (n° II HS), 135-162.

BRILLOUET J.M., CARRE B., 1983 - Composition of cell walls from cotyledons of Pisum sativum, Vicia faba and Glycine max. Phytochemistry, 22 (4), 841-847.

BRILLOUET J.M., RIOCHET D., 1983 - Cell wall polysaccharides and lignin in cotyledons and hulls of seeds from various lupine (Lupinus L.) species. J. Sci. Food Agric., 34, 861-868.

CHAMP M., BRILLOUET J.M., ROUAU X., 1986 - Non starchy polysaccharides of Phaseolus vulgaris, Lens esculenta and Cicer arietinum seeds. J. Agric. Food Chem., (accepted for publication).

COLONNA P., GALLANT D., MERCIER C., 1980 - Pisum sativum and Vicia faba carbohydrates : studies of fraction obtained after dry and wet protein extraction processes. J. Food Sci, 45 (6), 1629-1636.

COLONNA P., GUEGUEN J., MERCIER C., 1981 - Pilot scale preparation of starch and cell wall material from Pisum sativum and Vicia faba. Sci. Aliments, I (3), 415-426.

DAVIN A., BRILLOUET J.M., 1984. Séparation des protéines et des fibres à partir de tourteaux de lupin blanc et changeant par procédés humides. 2ème Congrès international du Lupin, La Rochelle (France), p.660-661.

FANTOZZI P., SENSIDONI A., 1983 - Protein extraction from tobacco leaves : technical, nutritional and agronomical aspects. In "Plant proteins for human food". BODWEEL and PETIT ed., W. JUNK publishers, 147-164.

FIORENTINI R., GALOPPINI C., 1983 - The proteins from leaves. In "Plant proteins for human food". BODWELL and PETIT ed., W. JUNK publishers, 131-146.

GUEGUEN J., 1983 - Legume seed protein extraction, processing and end product characteristics. In "Plant proteins for human food". BODWELL and PETIT ed., W. JUNK publishers, 63-99.

SANNELLA J.L., WHISTLER R.L., 1962 - Isolation and characterization of soybean hull hemicellulose B. Arch. Biochem. Biophys., 98, 116-119.

SOSULSKI F., YOUNGS C.G., 1979 - Yield and functional properties of air-classified protein and starch fractions from eight legume flours. J. Am. Oil Chem.Soc., 56, 292-295.

TRANCHINO L., COSTANTINO R., SODINI G., 1983 - Food grade oilseed protein processing : sunflower and rapeseed. In "Plant protein for human food", BODWELL and PETIT ed., W. JUNK publishers, 101-130.

TABLE I
COMPOSITION OF LEGUME SEED HULLS

SPECIES	YIELD[a],%	U[c]	POLYSACCHIDES-SUGARS[b], ANHYDROPOLYMERIC %						LIGNIN%	TOTAL[f,g] %
			RHA + FUC[d]	ARA	XYL	GAL	GLC[d,e]			
Kidney bean[1] Phaseolus vulgaris	6,2	16,6	(0.8)	10,5	10,8	1,6	44.1 (3.0)		1.2	85,5
Lentil Lens esculenta	4,5	10,6	(0.6)	3,9	9,5	1,6	37.4 (4.1)		1.7	54,7
Chick pea[1] Cicer arietinum	4,4	24,8	(2.5)	5,7	2,6	4,0	32.8 (3.8)		1.4	73,8
Lupin[2] Lupinus albus	20,0	8,6	0.4	6,9	17,6	1,5	49.6		2.1	86,7

1,4 : CHAMP et al, 1986 - 2 : BRILLOUET and RIOCHET, 1983

a : Percentage of hull in the seed. b : Mean from Saeman and TFA hydrolysis value. c : Uronic acids determined colorimetrically. d : (Rhamnose + Fucose) and glucose from TFA hydrolysis. e : Glucose from Saeman hydrolysis. f : Percent on dry weight basis. g : glucose from Saeman hydrolysis and rhamnose from TFA hydrolysis taken into account. Total does not include unidentified minor sugars.

TABLE II

COMPOSITION OF COTYLEDON CELL WALL POLYSACCHARIDES

SPECIES	YIELD[a], %	U[c]	RHA + FUC	ARA	XYL	GAL	GLC	TOTAL[e], %
			[d]					
Kindney bean[1]	10.7	0.28	0.08	1	0.21	0.09	0.37	63.2
		0.25	0.08	! 1	0.24	0.11	0.73	86.2
Lentil[1]	7.5	0.32	0.07	1	0.12	0.13	1.11	69.9
		0.26	0.03	1	0.14	0.11	1.25	87.3
Chick pea[1]	13.7	0.28	0.03	1	0.08	0.06	0.15	37.3
		0.21	0.03	1	0.11 !	0.08	0.42	56.7
Lupin[2] var. kalina	20.2	0.62	0.16	1	0.27	2.97	0.04	(70-80)
							0.33	
Smooth field pea[3]	6.9	0.36	0.06	1	0.11	0.10	0.29	81.9
							0.50	
Broad bean[3]	7.2	0.46	0.07	1	0.09	0.09	0.14	83.1
							0.40	
Soya bean[3]	9.3	1.04	0.25	1	0.29	1.60	0.07	72.7
							0.64	

POLYSACCHARIDES SUGARS [b]

1 : CHAMP et al, 1986 - 2 : BRILLOUET and RIOCHET, 1983 - 3 : BRILLOUET and CARRE, 1983.

a : Percentage of cotyledon dry matter - b : expressed as polysacchari-
dic material relative to arabinose (1) - c : uronic acids determined
colorimetrically - d : first line : TFA hydrolysis. Second line : Saeman hydrolysis - e : percent on dry weight basis - f : rhamnose and
glucose from TFA hydrolysis.

PROTEIN HYDROLYSATES

AND RUMEN FERMENTATIONS

A. Mordenti, R. Scipioni

Istituto di Zootecnia e Nutrizione animale
Università di Bologna
Via S.Giacomo 11, 40126 Bologna, Italy.

ABSTRACT

The administration of protein hydrolysates with a high content of free amino acids and peptides to animals of various zootechnical species induces growth promoting and probiotic effects not only ascribable to the plastic role of the amino acids. Results obtained in vitro and in vivo on beef and dairy cattle show certain modifications of the ruminal fermentations products (total VFA, acetate: propionate ratio, precipitable proteins, NH_3-N) probably indicating an increase in protein synthesis and in cellulosolytic activity.

INTRODUCTION

The efficiency in animal nutrition of small amounts of protein hydrolysates (peptones, peptides and free amino acids) is amply documented for pigs, where proteolysates play a role which can be defined as "extraproteic". Indeed, the results obtained illustrate: improved growth rates of the animals as well as improved digestibility of the nutrients, anti-stress effect, improvement of the reproductive efficiency and positive modification in the development of the intestinal microflora, sometimes accompanied by a reduction in pH. Promising results also were obtained with regard to the feeding of rabbits, veal calves and calves during weaning. In all cases, the effects observed can not be adequately explained by the composition of the proteolysates; therefore, the modifications in the development of the intestinal bacterial flora observed in pigs and in rabbits seem to be a more probable explanation.

RESULTS AND DISCUSSION

The results obtained in vivo are referred in tables I and II.

The research carried out on beef cattle (cross-bred young bulls of French breeds fed corn silage and siled sugar beet pulpes plus concentrate) has shown significant modifications of the net dres sing percentage at slaughtering and of the molar percent of rumen VFA (Table I). The second effect leads to the following interpre tation of the results: the pool of free amino acids and peptides favours the concentration of acetic acid, as also seen from other experiments performed _in vivo_ (on heifers) and _in vitro_ with ar- tificial rumens (Piva _et al_., 1986), even if this doesn't occur always, specially with proteolysates consisting of free amino a- cids only or with diets low in fibre; a reduction in protein de- gradation and an increase in the synthesis of microbial protein, to which the improved dressing percentage is probably ascribable, also was found.

The results obtained with the administration of small quan- tities of a proteolysate (free amino acids and oligopeptides) to high yielding dairy cows (Holstein Friesian) confirm the greater productive utility of this treatment for this category than for beef cattle, as the changes in ruminal fermentations observed in the latter suggested. In particular, the increase in acetate pro duction and the stimulation of the ruminal proteic biosynthesis seem the most reliable explanation, together with an increase in cellulosolytic activity, of the improvement in productivity of the earlier period when the peak was reached (from 5th to 4th week) and of greater persistence of the lactation curve.

REFERENCES

Mordenti A., Parisini P. and Scipioni R. 1984. New aspects in Ni trogen utilisation of Hy-Dairy Cows: researches on the use of free amino acids. Proceedings XIIIth World Congress Di- seases Cattle. Durban S.A. 17-21 sept.

Scipioni R. and Mordenti A. 1985. Effects of the addition of small amounts of pooled free amino acids from slaughter re- sidues to diets for beef cattle. In Feeding Value of by-pro ducts and their use by beef cattle. C.E.C. Agriculture Re- port EUR 8918 EN.

Piva et al. 1986. in press.

Table I - Main results obtained in beef cattle (Scipioni & Mordenti, 1983).

Groups		Control	Proteolysate (15 g/head/day)
No. of animals		18	18
No. of replications		3	3
Duration of trial	days	356	356
L.W. at the start	kg	193.75	193.08
Final live weight	"	593.55	595.05
Slaughter data:			
No. of animals		12	12
Slaughter live weight	kg	574.33	582.25
Dressing percentage (on empty l.w.)	%	67.34[B]	69.03[A]
Rumen fluid analyses:			
No. of samples		9	9
pH		7.59	7.53
C_2 : C_3 ratio		1.94[b]	2.22[a]

a, b = P<.05;   A, B = P<.01.

Table II - Main results obtained in dairy cattle (Mordenti et al. 1984)

Groups		Control	Proteolysate (8 g/head/day)
No. of cows		17	17
Milk production:			
Initial values			
(11 days after calving)	1	32.77	32.45
1st- 9th week of trial ([1])	"	36.58 (112)	37.56 (116)
10th-18th " " " "	"	34.54 (105)	34.85 (109)
19th-27th " " " "	"	29.86 (91)	31.25 (96)
Regression coefficient ([2])		.78[b]	1.22[a]
Milk quality:			
No. of samples		45	45
Fat	%	3.09	3.12
Protein	"	3.08	3.03
Urea	mg/kg	435.82	401.34
Acidity	SH/100 ml	6.88[b]	7.21[a]

a, b = P<.05.
([1]) Mean of individual weekly results.
([2]) Statistical analysis (RAO'S method) performed on all the individual results.

ORGANOSOLV TREATMENT OF WOOD AND ANNUAL PLANTS
- APPLICATION OF FINAL PRODUCTS -

J. Puls

Bundesforschungsanstalt für Forst- und Holzwirtschaft, Institut
für Holzchemie und chemische Technologie des Holzes
D-2050 Hamburg 80, Leuschnerstr. 91, FRG

ABSTRACT

A survey on organosolv activities in Europe is outlined.
There is a demand in long fiber pulp which can only be produced
by some of the organosolv processes presently under investigation. At the same time the market for high quality pulp is so
great that it cannot be served only by european producers.
Therefore organosolv pulps with characteristics of kraft pulps
have a good chance in the EEC. Organosolv lignins start to be
accepted by the chemical industry. Hemicelluloses from organosolv processes will not play an important role in the near
future. Organosolv processes which produce high quality pulp
and preserve the polymeric structure of hemicelluloses at the
same time are not yet within view.

INTRODUCTION

Presently wood pulping industry in the EEC has no problems
with respect to product utilization. Pulping processes, in
middle Europe mainly sulfite processes, have been developed
to a stage where remains of the high value pulp, the spent
sulfite liquor containing lignin-derivatives and hemicelluloses
are concentrated and burnt for recovery of the pulping chemicals. The energy produced may cover the demand of the pulp
factory as well as of a integrated paper factory. Completely
different is the situation of the remaining pulp factories
based on straw. Chemical pulping processes based on straw use
alkaline cooking liquors. The chemical recovery shows an important difference. Straw contains 4-6 % silica as compared to
less than 1 % in wood. The silica dissolves in the black liquor
as silicate, which causes serious problems during evaporation.
Therefore traditional straw pulping units are open processes
with increasing environmental problems. There exist an
urgent demand for improvement of pulping processes for
annual plants. One process recently been developed is the

swedish NACO process which introduces oxygen into an alkaline pulping liquor (Hartler and Ryrberg, 1985). Even organosolv processes may have the potential for producing pulp from straw.

ORGANOSOLV PROCESSES IN EUROPE

The pulping industry in middle Europe is interested in organosolv pulping for different reasons:
1. Organosolv pulping offers the chance to operate economically small-sized pulp mills which may precisely cover the need of one paper factory. Modern sulfite mills however need such a big production capacity, that capital costs and raw material supply become critical factors.
2. Organosolv pulping offers a broader raw material basis, because wood species may serve as raw material, which cannot be pulped in sulfite plants e.g. Pinus species.
A complete overview of world-wide organosolv activities cannot be given here because sulfur-free techniques which use organic solvents to separate the lignin and the carbohydrate portion of wood include aqueous phenol, aqueous methanol, ethanol, butanol, ethylen glycol, formic acid, acetic acid under additon of various catalysts of different quantities.
A critical comment on some of these processes was recently given in Pulp & Paper Canada (1984). Moreover organosolv processes have not only been developed as pulping processes but also as saccharification processes (Paszner, 1984) and for pretreatment of lignified material for further biotechnical conversion (Wang and Avgerinos, 1983, Holtzapfel and Humphrey, 1984). Cost analysis of different pretreatment options revealed organosolv processes to be expensive in comparison to other pretreatment processes (Fan et al., 1982). Therefore in Europe interests in organosolv processes are mainly focussed on pulp production. Major european pulp and/or paper producers as well as research organizations have manifested their interests in organosolv delignification through patents and publications (Table I). Early work utilized aqueous solutions of ethanol to remove lignin and to produce wood pulp (Kleinert and Tayenthal, 1931). More recently n-butanol among

the primary alcohols and ethylene glycol among the polyhydroxy
alcohols were the most promising candidates for producing a
well-pulped residue and for lignin removal (Gast et al.,1983).

Careful experiments with ^{13}C-labelled ethylene glycol
revealed some percentages of this pulping chemical to be
chemically bound to the lignin. Economics however require a
solvent recovery of nearly 100%. This is valid for all alcohols. This requirement is nearly impossible to fullfil,
therefore alcohol pulping was reported to be only economically feasible if by-products (lignin and hemicelluloses)
can be upgraded and sold (Katzen et al., 1980). In practise
organosolv systems with no added catalyst turned out to
develope too little delignifying capacity, especially for
softwoods. Therefore most systems need the addition of catalysts, which is reflected in Table I. Addition of catalysts
however has consequences on the properties of the end products.

A typical acid-catalyzed process is the Batelle-Geneva
process (Johansson et al., 1983). The delignification of lignocelluloses in the presence of phenol and hydrochloric acid
results in the hydrolysis of hemicelluloses into monomeric
sugars and production of cellulose pulp. Apparently the pulp
is affected with respect to degree of polymerisation ($\overline{DP}_w$)
under the action of the catalyst. It seems only to be usable
for certain chemical pulp qualities. Cellulose ether manu-
facturers however need pulp qualities of high $\overline{DP}_w$. Their
european raw material basis is becoming smaller and smaller.
Increasing numbers of traditional sulfite mills introduce new
bleaching chemicals into their bleaching sequences in order
to substitute chlorine. The new bleaching chemicals did however
attack the cellulose chain to higher degree than chlorine.
As a consequence nearly all pulp and paper producers favour
base-catalyzed systems (see Table I).
The MD-Munich process (Edel, 1984), as a typical process of
this category, produces a bleached pulp with strength properties comparable to long-fibre sulfate pulp. Addition of more
alkali than catalytic amounts has an effect on lignin- and
hemicellulose quality. The MD-Munich process as a two-step

process recovers part of the hemicelluloses in a polymeric form during the first step (methanol : water cook). Most of the lignin and part of the hemicelluloses are solubilized during the second step. It is unavoidable that the hemicellulose part is degraded under strong alkaline conditions at elevated temperatures. With respect to lignin NaOH-additon may result in a higher ash content.

Special features of each process are reflected in the chemical compositon, functional groups, molecular weight and other physical parameters of hemicelluloses and lignin, which are of decisive importance for their further utilization. Another aspect which has turned out to be of tremendous importance for producers of these new products is the availability of the new products and reproducibility of their chemical and physical characteristics. Most representatives of the chemical industry were unwilling to think about organosolv lignins as a chemical raw material, because it was only available in the 100 g-scale. On the other hand the pulping industry was shy to invest in investigations of organosolv processes, because the need for lignin and hemicellulose by-product utilization was obvious. The products however were not in demand. In Germany it is partly the merit of MD-Organocell in Munich to operate a pilot plant supplying possible customers with enough lignin for their specific applications. Operation of the pilot plant is supported by the BMFT and EEC DG XII.

UTILIZATION OF ORGANOSOLV LIGNIN

Two major chemical companies have investigated possible applications for organosolv lignins. Bayer AG has investigated the transformation of ethanol : water lignins to polyurethan foams, which were comparable to commercial sucrose-started hardfoampolyethers with respect to functionality and OH-number. The only objection was the dark colour of the product in comparison to colourless sucrose-based polyethers. However the company only felt able to judge the commercialisation of the product after the organosolv process had entered production scale itself (Niederdellmann, 1982). BMFT published a study of

Ruetgerswerke on the utilization of organosolv lignin and
hydrolysis lignin as a chemical raw material (1984) with
regard to

- Thermolysis of organosolv lignin in the presence of 9.10-di-
  hydro-anthracene or coal tar fractions (coal tar pitch, aromatic oils) at 300 - 400°C. Phenols and phenol homologues
  were formed in total yields up to 18 % per weight, related
  to lignin input, and also a thermoplastic residue consisting of lignin oligomers (formed by lignin depolymeri-
  sation) and aromatic hydrocarbons from the tar fraction used.
- Utilization of lignin in phenol-formaldehyd resins. In
  particleboard resins approximately 10 % of the resin could
  be substituted by organosolv lignin without violating the
  normalized characteristics of chip boards of highest quality.

In conclusion Ruetgerswerke stresses the model character of
the investigations. More detailed studies on lignin hydration
with dihydro-anthracene were suggested. Both studies as well
as investigations of Beinhoff (1986) confirm the literature
data on organosolv lignins, i.e. the exhibition of much lower
average molecular weight than kraft lignin and similar methoxyl
values as milled wood lignin (Table II).

Both studies show that lignin is gaining acceptance as
a chemical raw material in industry. Its introduction seems
to be a question of its price compared to the oil price. Both
studies reflect possible large quantity utilization of lignin
as a polymeric product and after conversion to low molecular
weight chemicals.

Additionally, investigations are being pushed forward to
use lignin in small volumes, but for high value specific
purposes, as for pharmaceuticals (Mach, 1981), controlled
release agents (Tinnemans et al., 1985), specific adsorbants,
antioxidants and reinforcing agents. It should also be mentioned that recent progress in biotechnology may offer ex-
citing possibilities for direct changes of lignin in terms
of chemical structure and physical properties.

UTILIZATION OF HEMICELLULOSES

Utilization of hemicelluloses derived from organosolv process aredepend upon the raw material and the nature of the catalyst used. Softwood hemicelluloses consist of a mixture of hexosans and pentosans, whereas the hemicellulose fraction of hardwoods and annual plants is nearly uniform, yielding pentosans. Acid catalyzed processes result in a degradation of the polymers into monomers. Some sugar degradation products may already occur although the hydrolysis process is clearly slower in mixed organic-aqueous media than in pure water. In organosolv systems with no added catalyst the hemicelluloses are recovered as a mixture of oligomeric and polymeric carbo-hydrates similar to the hemicellulose fraction after steaming-pretreatment (Puls et al., 1985). Base-catalyzed organosolv systems will degrade hemicelluloses under formation of saccha-rinic acids and numerous fragmentation products. With regard to acidcatalyzed processes reaction of hemicelluloses may not stop in the monosaccharide state, but may be transformed further to furfural and hydroxymethylfurfural. Since these compounds have to compete with petrochemicals, production costs must be low. Furfural once served as a feedstock for nylon but was substituted by oil-derived butadiene. It should be mentioned that hexoses can be fermented to ethanol whereas pentoses can be converted to single cell protein, a practice which had been exercised in Germany up to the 1950's. After World War II imported soya became cheaper. Nowadays only Cellulosefabrik Attisholz in Switzerland uses these old tech-niques in an updated form.

Recently yeasts, fungi and bacteria have been identified for the conversion of pentoses into ethanol and other organic solvents. However strains have not yet been developed which may be used in industrial processes (Fiechter, 1983). Organo-solv processes without adding catalysts offer the possibility to recover polymeric hemicelluloses as comparatively pure fractions. Hardwood xylans could be derivatized or used as such for very special, high value applications (Tables III and IV). Xylitol production is another profitable five carbon sugar utilization. Xylitol possesses certain properties that could make it valuable commercially. Besides of furfural xylitol is the only product on the market up to now made of hemicelluloses. Its distribution however is limited because it is only allowed to be used as food additive.

These possible hemicellulose utilizations are not specific for fractions of this component from organosolv processes. They are also valid for component fractionation principles like steaming, viscose- and prehydrolysis processes. The unique properties of lignin derived from organosolv treatment is only achievable however by pulping lignocellulosics with organic solvents. A broader application of lignin as a renewable resource is desirable.

REFERENCES

Hartler, N. and Ryrberg, G. 1985. Comparison between utilization of cellulose for paper from wood and straw. In Hill,R.D. and Munck, L. (ed.). New Approaches to Research on Cereal Carbohydrates. Elsevier Science Publishers B.V., Amsterdam, 323-327.

Anonymus. 1984. Organosolv pulping processes-boon or boondoggle? Pulp and Paper Canada,85,No.7,15-17.

Paszner, L. 1984. Organosolv process for hydrolytic decomposition of lignocellulosic and starch material. PCT Int. Appl.WO 84 03, 304.

Wang, D.I.C. and Avgerinos, G.C. 1983. Selective solvent extraction of cellulosic material. U.S. US 4,395-543.

Holtzapple, M.T. and Humphrey, A.E. 1984. Effect of organosolv pretreatment on the enzymatic hydrolysis of poplar. Biotechnol.Bioeng.,26, 670-676.

Fan, L.T., Lee, Y.-H. and Gharpuray, M.M. 1982. The nature of lignocellulosics and their pretreatment for enzymatic hydrolysis. Adv.Biochem.Eng.,23, 155-187.

Edel, E. 1984. Das MD-Organosolv-Zellstoffverfahren. Deutsche Papierwirtschaft-dpw.,1, 39-45.

Kleinert, T. and v.Tayenthal, K. 1931. Über neuere Versuche zur Trennung von Cellulose und Inkrusten verschiedener Hölzer. Ztschr. ang. Chem.14, 788-791.

Gast, D., Ayla, Ch. and Puls, J. 1983. Component separation of lignocelluloses by organosolv treatment. In Strub,A., Chartier,P. and Schleser,G. (ed.). Energy from Biomass. 2nd. E.C.Conference, Appl.Sci.Publishers, London and New York, 879-881.

Katzen, R., Frederickson, R. and Brush, B.F. 1980. The alcohol pulping and recovery process. Chem.Eng.Prog.,76,62-67.

Johansson, A., Sachetto, J.P. and Roman, A. 1983. A new process for the fractionation of biomass. In Strub,A., Chartier, P. and Schleser,G.(ed.). Energy from Biomass. 2nd. Con-

ference, Appl.Sci.Publishers, London and New York,
868-872.

Niederdellmann, G. 1982. Versuche zur Gewinnung von Hart-
schaumpolyurethan aus Hydroxylgruppen-haltigen Bestand-
teilen der Organosolv-Aufschlußlösung. Kolloquium Organo-
solv-Aufschluß lignocellulosischer Rohstoffe. March 26th,
BFH Hamburg.

Ruetgerswerke AG, 1984. Studie zur Nutzung von Lignin als
Rohstoff für Chemieprodukte. Final Report BMFT - FB -
T 84 - 199, 54 pages.

Mach, W. 1981. Neuer biochemischer Wrkstoff, Verfahren zu
seiner Herstellung und diesen Wirkstoff enthaltende
pharmazeutische Zubereitung. Eur.Pat.Appl. 8010 1275.8.

Tinnemans, A.H.A., Martens, H.F., van Veldhuizen, G.J. and
Greidanus, P.J. 1985. Chemically modified lignins for
use in controlled-release devices. Proc. I.S.W.P.C.
Vancouver,B.C. August 26-30, 105-107.

Puls, J., Poutanen, K., Körner, H.-U. and Viikari, L. 1985.
Biotechnical utilization of wood carbohydrates after
steaming-pretreatment. Appl.Microbiol.Biotechnol.,22,
416-423.

Fiechter, A. 1983. Pentoses and Lignin. Advan.Biochem.Eng./
Biotechnol.,27, 186 pages. Springer Verlag, Berlin
Heidelberg New York Tokyo.

Beinhoff, O. 1986. Unpublished results.

Table I : Patents and Publications reflecting Interests of European Companies and Individuals
in Organosolv Processes

Literatur	Company	Solvent	Catalyst
EUR 8641, Prod.Feed.Single Cell Protein, 38-49	Cellul.Attisholz	MeOH : H_2O	NaOH
Ger.Offen. DE 3, 212, 767	MD - München	MeOH : H_2O	NaOH : Anthrachinon
PCT Int.Appl. WO 8302, 125	Battelle Memorial Institute	PhOH : H_2O	HCl
Eur. Pat. Appl. 8, 783	Benckiser-Knapsack	EtOH : H_2O	H_2SO_4
Austrian Appl. 4, 859/81	Chemiefaser Lenzing	EtOH : H_2O	
PCT Int. Appl. WO 84 03, 527	Nivelleau de la Bruniere and Galichon	ethylene glycol : H_2O	NaOH
Neth. Appl. NL 82 OO, 241	Van Elten Engineering	MeOH : H_2O	H_2SO_4 : $MgCl_2$

Table II : Organosolv Lignins in Comparison to Natural (Milled Wood) Lignins

	Molecular Weight $\overline{M}_w$	% Methoxyl
Milled Wood Lignin Spruce	12.150	16.9
MD-Me OH : H_2O Lignin Spruce	7.920	14.5
Milled Wood Lignin Beech	20.800	22.3
BFH-ethylene glycol : H_2O Lignin Birch	6.120	18.0

Table III: Derivatives of Xylans

Derivative	Application	Literature
Carboxylmethylxylans	Detergents	Schmorak,J. and Adams,G.A. TAPPI 40, 378 (1957)
Michael addition with acrylonitrile	Thickener of organic solvents	Nordgren,R. Ger.Offen. 2, 064, 810(1971)
Acetates	(e.g. Higher	Carson,J.F. and Maclay,W.D.
Butyrates	(fatty acid	J.Am.Chem.Soc.,68,1015(1946)
Benzoates	(extenders	Husemann,E.;J.Pract.Chem. 155,13(1940)
Reaction with epichlorohydrin	Flocculents and adhesives	U.S.Patent 3, 833, 527(1974)
Polysulfate esters	Clinical application	Raynaud,R. et al. Therapie 20, 1259(1965)
Xylan sulfates	Antithrombotic and hypolipemic activity	FR 2, 543, 145 (1984)

Table IV: Utilization of Polymeric Hemicelluloses

4-O-Me-Glucuronoxylan	Antitumor effect	Hashi,M. and Takeshita,T. Agr.Biol.Chem.1979,43,951-9
4-O-Me-Glucuronoxylan	Water-absorbing agent	Ger.Offen.DE 3, 405, 208
Glucomannan	Caviar substitute	Jpn. Tokkyo Koko JP 8146,751
4-O-Me-Glucuronoxylan	Extender in PF-resins	Ger.Offen.DE 3, 213, 159

CELLULOSE DEGRADATION

Concentrated acid hydrolysis processes;
application of the final products

Hans M. Deger
Hoechst AG, D-6230 Frankfurt 80
FRG

Summary

Acidic cellulosehydrolysis for dextrose-production seems not to be an
alternative to starch based processes, excluded areas, where starch is
short in supply. The economics of such processes are depending on the
options for added value products from hemicellulose and lignin. There
are some applications for these products, but further research is necessary to develop a broad range of application and market-opportunities.

1. Introduction

Cellulose and starch are naturally occuring polymers consisting of
dextrose as the monomeric building block. The major difference between the two carbohydrate polymers arises from the type of bonding:
in starch the dextrose units are linked by α-1.4 and α-1.6-bondings,
whereas cellulose consist only of β-1,4 linked dextrose. The tertiary
structure of starch is therefore a wide helix disturbed by branching
points, whereas the cellulose helix is very narrow and gives rise to a
high degree of cristallinity. In nature starch occures in plants forming a compound with proteins, cellulose commonly occurres as a compound with lignin and hemicelluloses. Because of this, the discussion
of cellulose-degradation is strictly connected with its competition
with starch-hydrolysis, separation-techniques and added value-products
derived from the byproducts hemicellulose and lignin.

2. Pretreatment

Pretreatment is usually performed by treating the basic biomass with
dilute acid at about 100 °C. The hemicellulose is cleaved into monomeric sugars, the cellulose/lignin-compound is seperated by a simple
filtration/washing step from the soluble hemicellulose-sugarmixture.
Newer processes, e. g. steaming yield soluble oligomers, which probably
are a basis for new added-value products.

3. Concentrated acid hydrolysis-processes

There are two major routes to acidic cellulose hydrolysis

- dilute acid, high temperature processes
- concentrated acid, low temperature processes

Dilute acid processes are limited by the similar kinetics of cellulose cleavage and sugar decomposition under the conditions required and therefore usually give low yield and a variety of byproducts which at least hamper the isolation of pure dextrose. In these processes the acidic agent is usually not recycled.

In concentrated acid processes yields are usually high (90 %), the need for a posthydrolysis step (for dextrose production) and the process-design for acid-recycling is limiting for the economical feasibility of such processes.

In the last three decades there were several pilot projects reported in literature using conc. sulphuric acid, superconc. hydrochloric acid and recently the IG-Farben HF-process has been reinvestigated by HOECHST AG.

Noguchi process

In the Noguchi-Process concentrated H_2SO_4 has been used for hydrolysis of lignocellulosic-biomass at temperatures below roomtemperature and membranes were used for carbohydrate seperation. A pilot-plant was built in the early sixties and closed down again after less than 6 mounth of operation due to unsolvable problems in the membrane reactor.

Rheinau process

The Rheinau process used superconcentrated HCl for cellulose-hydrolysis. The acid was recovered by distillation, the excess HCl-gas adsorbed to $CaCl_2$-solution and recycled to the hydrolysis-vessel by heating. A pilot-plant at Rheinau near Ludwigshafen/FRG has been in operation for several years. Obviosly the process was limited by the enormous corrosion problems due to the aggressive media required.

Some years ago ICI reported a new approach using salt-catalysts to support hydrolysis in HCl. The overall design of this process should be very similar to the Rheinau-Process.

Hoechst process

In recent years Hoechst AG reinvestigated a process published in the early thirties by IG-Farben using dry hydrogenfluoride-gas. The major advantages of this approach are:

- dry HF-gas is not corrosive
- no water is introduced to the process
- HF can be easily recycled by heating

The process is designed as follows (Fig 1):

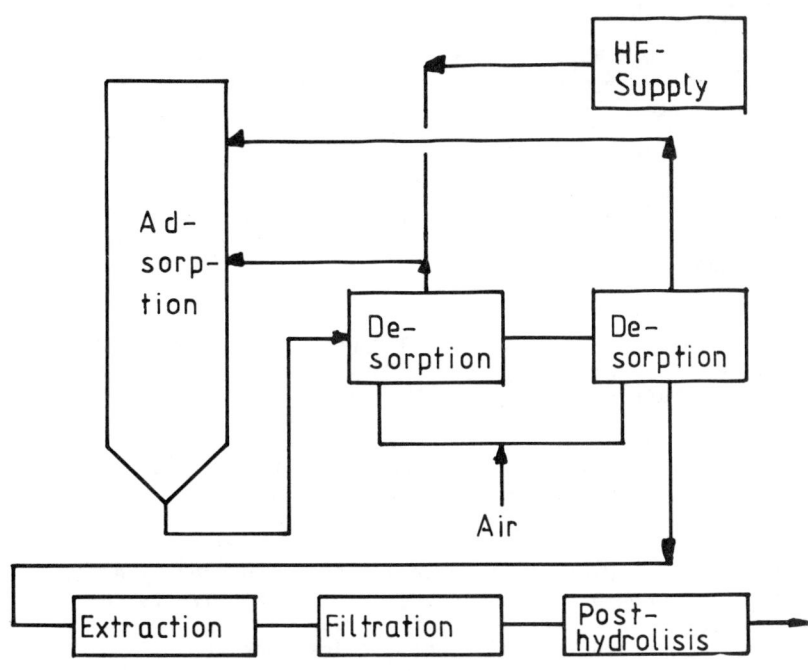

Fig. 1)   Scheme of Hoechst HF-Process

Wood chips, dried to a water-content of about 10 %, are continously fed into an absorbtion-tower from above and HF-gas is introduced until a concentration of 40 - 60 % w/w is reached. After a dwelltime of 30 - 60 min the
HF-containing biomass, still solid and appearing dry, is transferred into
desorption units, where under support of heat and air, the HF is withdrawn
until a final concentration of 0.1 - 0.2 % of HF is reached. The desorbed gas
streams are cooled and recycled to the absorber. Sugar is seperated from the
residual solids by hot-waterextraction, seperated from the insoluble lignin
and posthydrolysed to yield a monomeric sugarsolution of a concentration of
10 - 15 %. If prehydrolysed lignocellulose is used a almost pure Dextrose
solution is obtained, using basic biomass a mixture of C6 and C5-sugars is
harvested. The yield of Dextrose is more than 90 %, pentose-yield about 80 %.
The crucial steps of the process (absorption and desorption) have been tested
in a small pilot unit and no prohibitive technical objections could be found.

Calculations have proofed, that the process is energysufficient for ethanol-production when lignin is used for energy supply. However, compared with starch-based ethanol-processes, the process is unlikely to be economic due to high feedstock prices and a lack in added value-products for the hemicellulose- and lignin-fraction.

4. Chemicals from hemicellulose

The development of added value-products from hemicellulose is limited by the uncertain feedstock and because of the unstability of C_5-sugars compared with C_6-carbohydrates. There are two hemicellulose-based products on the market:

- Furfural:
  Being produced from grain-waste (oat-hulls, corn-cops), Furfural is about twice as expensive as Phenol, with which it competes directly in its high-tonnage applications (resins).

- Xylitol:
  Xylitol has been launched as a sweetener with promising properties. A broad introduction of the product is delayed due to unfavorable toxicological findings, which need further experimental work and explanation.

To find a broad use for hemicellulose, chemistry of pentoses must be further investigated and new applications for pentose-derivatives must be found.

The isolation of oligosaccharides from hemicellulose, produced by e. g. steamingprocesses could probably be an interesting alternative to the products mentioned above. In further chemical derivatisation, the oligosaccharides could be converted to surfactants, thickeners etc.

5. Chemicals from lignin

Despite the good properties of reactive lignin (e. g. from Rheinau Process) in resins, all efforts in producing pure chemicals from lingnin failed. There where good lab-scale results e. g. in reductive deavage, but scale-up always failed due to severe problems in separating the mixtures obtained. Using the new separation techniques developed in the last decades, those problems could probably be overcome, but in terms of economics lignin is not a promising feedstock for bulk chemicals.

THE FRACTIONATION OF LIGNOCELLULOSICS
FOR THE PRODUCTION OF CHEMICALS

J.-P. Sachetto, J.-M. Armanet, A. Roman[*] and A. Johansson[**]

BATTELLE, GENEVA RESEARCH CENTERS
7, route de Drize
CH - 1227 Carouge-Geneva

* Present address: Biogen SA, 46 route des Acacias,
CH - 1227 Carouge/Geneva
** Present address: Technical Research Center of Finland (VTT)
Vuorimiehintie 5, SF - 02150 Espoo

ABSTRACT

Among the different routes for the up-grading of ligno-cellulosics
into chemicals, "refining" or separation of the three major constituents is
a promising one. The central idea developed by Battelle-Geneva was to use
a solvent by which the lignin can be extracted leaving the cellulose
untouched while the hemicellulose is converted into an aqueous pentose solution. The lignin can be further processed into phenol by hydrocracking, into
resins as a phenol substitute or into specialty chemicals through derivati-
zation. The pentoses can be dehydrated into furfural or fermented into SCP.
The cellulose can be used for its polymeric properties but also converted
efficiently into glucose by means of a convenient saccharification process.

INTRODUCTION

During the last years, an intensive effort has been directed towards
the development of new processes for the up-grading of biomass. Still to
date, however, the only economically viable method for the utilization of
ligno-cellulosic material is the manufacturing of cellulose pulp for paper
production. In this, only the fibrous part of the ligno-cellulosic material
is recovered whereas the dissolved lignin fraction after concentration is
burnt as fuel to support the energy requirements of the pulping process.
In fact, much of the energy produced by burning the lignin is consumed in
the concentration of the lignin bearing liquid, and thus only a fraction of
the total energy content of the residues is available as net energy.

This latter fact seems also to be true for most of the hydrolysis
processes suggested to date. In these, the fibrous cellulosic fraction is
depolymerized into its sugar building blocks through digestion with acids,
and the lignin part is suspended in an aqueous phase which has to be dried
before burning; thus very little or none of its fuel value is recovered as

net energy.

Today, when one speaks about the up-grading of biomass, one is generally referring to the quantities of ligno-cellulosic material not used as raw materials for the pulp or building industry. Because of fierce competition, the quality requirements for the pulps produced and hence the materials required for pulp mills are extremely stringent, (surpassing in many cases actual needs required in practice for the final products (Leopold, 1982). Furthermore, present day pulping units are immense and thus the use of odd amounts of biomass of varying quantity or in unfavorable geographical position is not economically justified.

Hence it is for these residual amounts, composed of mixed agricultural and forest residues, that new processes are sought. Such processes should not be very sensitive to scale-up, allowing relatively small units to be constructed in areas where the local supply of raw material or demand for specialty pulps is sufficiently dense avoiding transport of bulky material. For an economically viable operation it is also necessary that the lignin fraction can be up-graded in a reasonably efficient way. Ideally, the process should be such that all three of the main constituents of ligno-cellulosic materials, cellulose, hemi-cellulose and lignin, can be separated and up-graded independently, without the use of expensive chemicals and large amounts of energy.

THE FRACTIONATION PROCESS

The basic idea of this new approach developed at Battelle-Geneva is to use a solvent by which the lignin in the lignocellulosic material can be extracted leaving the cellulosic fraction untouched.

In the past and up to recently different solvents have been investigated but they require high temperatures and pressure to be efficient. They are also generally difficult to recycle. As the process normally has to be carried out at elevated temperatures the solvents being volatile result in process losses too significant to be acceptable.

It has been known for long that phenols are good solvents for lignin in particular in the presence of acid catalysts (Brauns, 1952). However, it is also known that when using phenol as lignin solvent under acidic conditions the phenol is rapidly lost through reaction with liquid and

possibly with the furfural formed from the hemicellulose present in the raw
material (Schweers, 1974, April, 1979).

In Battelle-Geneva process, phenol is used as delignification solvent
but the resinification is avoided. The delignification is carried out at
100°C and atmospheric pressure. Under these conditions, the phenol is
totally miscible with the water phase forming a homogeneous liquid phase
into which the lignin is dissolved while the hemi-cellulosic fraction is
hydrolysed simultaneously.

The remaining fibrous cellulose fraction is simply separated by
filtration and the liquid phase is allowed to cool. During cooling, the
phenolic portion with the dissolved lignin separates spontaneously from the
aqueous phase containing the pentoses.

The aqueous phase can be recirculated in order to increase the concentration to a desired level before further treatment of the pentoses. The
phenol is separated from the hydrophobic fraction and recirculated in the
process.

The remaining lignin fraction which is free of sugars can be recovered
for further processing.

The overall scheme is best visualized by the simplified schematic
flow-sheet in Fig. 1.

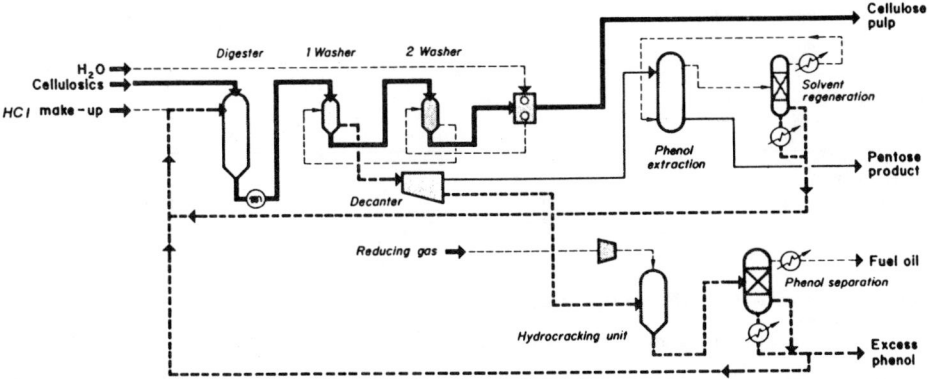

A prehydrolysis of the hemicellulose and a dissolution of the lignin
takes place in the digester. The fibrous cellulose fraction is separated
and washed in standard type pulp washing equipment. The pentose containing
aqueous phase is separated from the hydrophobic lignin solution in the

decanter.

This latter can be treated in a hydrocracking unit to transform the dissolved lignin into phenol and fuel oil. Part of the phenol can be used as make-up in the process to cover technical losses while the excess constitues a valuable by-product of the process.

The aqueous phase is treated by solvent extraction in order to separate the pentoses and the phenol dissolved in the aqueous solution.

Typical results obtained in laboratory experiments with different raw materials are presented in Table I.

TABLE I – LIGNOCELLULOSE REFINING: TYPICAL RESULTS ON DIFFERENT RAW MATERIALS

Raw materials:	Birch	Bagasse	Red Oak	Wheat Straw	Spruce
Pulp yield (dry basis %)	40	40	35	38	43
Residual lignin on pulp (%)	1.2	2.5	4	3.2	7.6
Lignin dissolved (%)	97.7	95.5	94	92.5	88.6
Average DP of pulp (CuEN)	1,500–2,000	500–900	–	450–900	2,000
Carbohydrates in pulp (%) . Pentosans . Cellulose	5.3 92	4.5 90	5.3 87	3.5 90	0.5 91
Yields pentosans hydrolysis (%)	92	93	91	95	99

THE UP-GRADING OF THE THREE FRACTIONS

The fractionation process leads to three individual streams which can be processed and up-graded separately. Non-exhaustive exemples of the treatments which can be applied and their corresponding end-products are given in Table II.

TABLE II – UP-GRADING OF THE THREE MAJOR FRACTIONS OF THE LIGNOCELLULOSICS

Component	Processing	End-products
Soluble sugars (mainly pentoses)	– heat, pressure – catalytic hydro-genation – fermentation	– furfural, furanic resins – xylitol, polyols – SCP, amino-acids
Cellulose	– defibrillation – dissolution, regeneration – derivatives – saccharification	– paper, fluff – viscose – CMC, cellulose ethers and esters – glucose, methylglucose, ethanol, fermentation products
Lignin	– hydrocracking – reaction with formaldehyde – derivatives	– phenol, fuel oil – phenolic type resins – specialty polymers (pre-polymers, surfactants)

THE SACCHARIFICATION PROCESS OF THE CELLULOSE FRACTION

The crude wet cellulose pulp obtained through the fractionation of the ligno-cellulosic material is a rather convenient feedstock for saccharification. Battelle-Geneva has developed a new acidic saccharification approach for this type of feedstock which combines the advantages of both the dilute acid and concentrated acid processes.

In this approach the operating conditions (temperature, pressure) and the yields of glucose are those of the concentrated acid hydrolysis methods, whereas the acid consumption per ton of sugar produced is similar to that of the dilute acid hydrolysis methods.

When cellulosic feedstocks with a low lignin content are available (e.g. cellulose from delignification processes, waste paper, waste fibres, etc.), the saccharification problems are simplified because the diffusion of the liquid acid catalyst within the cellulose fibres is much easier. Further, the amount of insoluble lignin which appears in the hydrolyzate

becomes negligible. Consequently less liquid acid is required for the hydrolysis of a given amount of cellulose and the volume of acid to be further
treated per ton of sugars is lowered.

The new Battelle-Geneva hydrolysis concept involves the use of concentrated hydrochloric acid in such a way that the amount of acid consumed per
ton of glucose produced is minimized still maintaining high glucose yields
(generally above 90% of the potential glucose content of the feedstock).

A comparative acid-sugar balance for several industrial acidic saccharification processes is presented in Table III, together with the figures
for Battelle-Geneva hydrolysis process as determined from bench-scale pilot
experiments.

From Table III it can be seen that the acid consumption of the new
Battelle process is substantially lower in absence of total acid recovery
compared with the other processes, resulting in a decisive improvement in
this saccharification process in reducing extensive energy consumption and
capital cost normally related with the acid recovery of conventionnal concentrated acid processes (Bergius type).

The cellulose pulp used in the bench-scale pilot experiments was issued
from the ligno-cellulose fractionation process. In this, the residual lignin
content was below 4%. The process tolerates a water content of the feedstock
of up to 40%. Above this value it is preferable to remove the excess
humidity.

Waste newspaper fibres provide also glucose with high conversion yields.
The same applies to crude waste fibres from pulp mills, which generally are
discarded owing to their poor physical properties and can amount up to
5,000 t/year on a pulp and paper site.

A major objective of the process development programme has been to
adapt this saccharification technique to existing industrial equipement for
facilitating scale-up and avoiding both technical uncertainties and costly
equipement development phases.

Corrosion problems can be overcome using existing materials of
construction.

The attractiveness of such a process lies in the low acid consumption,
the absence of a complicated recovery unit and the fact that glucose is

TABLE III - COMPARATIVE ACID-SUGAR BALANCE OF VARIOUS ACID SACCHARIFICATION PROCESSES INCLUDING THE BATTELLE-GENEVA PROCESS

TYPE OF PROCESS	DILUTE SULFURIC ACID PROCESS (Inventa AG, 1981)	LIQUID 40% HCl BERGIUS RHEINAU PROCESS (U.S. Patent, 1962)	CHISSO GAZEOUS HCl PROCESS (J. Kusawa, 1966)	BATTELLE-GENEVA CONCENTRATED PROCESS (Eur.Patent Appl. 0093 088)
RAW MATERIAL	Total wood	Total wood	Pre-hydrolyzed wood	Cellulose-enriched fibres
SUGAR CONCENTRATION IN ACID	3-4% in 0.8% H_2SO_4	20% (after recycling) in 40% HCl	29% in 37% HCl	54.3% in 43% HCl
SUGAR CONCENTRATION IN ACID AFTER PARTIAL ACID RECOVERY	-	24% in 24% HCl	33% in 24% HCl	70.5% in azeotropic HCl solution
ACID CONSUMPTION PER TON OF SUGARS IN ABSENCE OF TOTAL ACID RECOVERY	0.25-0.30 t	0.76 t	0.49 t	0.10 t

recovered in a concentrated aqueous stream, ready to be used for further processing.

All this makes it possible to envisage this process also for converting rather small amounts of cellulose wastes available on industrial sites or in urban areas, without having the economic disadvantages inherent in small-scale operations.

CONCLUSION

The fractionation of ligno-cellulosics appears to be a suitable approach for an integrated up-grading of this feestock.

Altogether the cellulose, the soluble sugars and the lignin obtained by the Battelle-Geneva process make up to 90% of the starting raw material. Each of these separated streams offers valuable outlets. The advantage of the process developed by Battelle is that it is feasible on a much lower scale than the conventional wood pulping processes. This is due to milder processing conditions, an easier recovery of the chemical reagents (no recovery boiler needed) and a larger material output. Ligno-cellulosics constitute a potential source of glucose, a substrate for fermentations. The cellulose pulp obtained in the Battelle-Geneva fractionation process can be converted very efficiently into glucose by means of a gaseous hydrogen chloride hydrolysis.

Technically, the above processes (fractionation and saccharification) appear feasible. Their economical feasibility will depend upon the industrial environment and must be assessed on a case by case basis.

REFERENCES

April, G.C. et al, 1979. Tappi, 62, No.5, 83.
Brauns, F.E., 1952. "The Chemistry of Lignin", Academic Press, 488.
European Patent Application, 0093 088, 1983.
Inventa AG, 1981. Chemical Engineering, June 15, 62-65.
Kusama, J., 1966. J. Chem. Soc. Jap, Industr. Chem. Sect., 69 No.3, 469-489.
Leopold, B., 1982. Tappi, 65, No.2, 11.
Schweers, V.H.M., 1974. Phenol Pulping. Chemtech, August, 490.
U.S. Patent 3.067.065, 1962.

Shuttle HCl process for the preparation of oligosaccharides containing
products from wood
(J.P.M. Sanders° and P.E. Linnett[+])

° Gist brocades R&D, Delft, the Netherlands
[+] Shell Research, Sittingbourne, England

Comminuted biomass which comprises as main components cellulose, hemi
cellulose and lignin and which may be derived from comminuted trees
(hard wood and soft wood), plants, grasses and waste materials has long
been recognised as a source of useful carbohydrates such as cellulose
and sugars.
A great deal of research effort is currently being spent in this field
of technology with the aim of producing sugars, animal feed and liquid
fuels like ethanol from biomass.
Physical pretreatments such as milling of different sort, various steam
treatments, solubilization pretreatments using metal ion solvents like
$ZnCl_2$ and Cadoxen nor chemical pretreatments such as dilute sulphuric
acid and concentrated HCl or HF treatments meet the very severe crite
ria as set by the combination of economics, application and environ
ment.
The physical pretreatments do require a major cellulolytic step in
order to valorize the cellulose for animal feed or for the carbohydrate
need of microorganisms. Apart from the waste streams they produce, the
use of chemicals is more often prevented by their mere costs.
As a result of extensive experimentation, we have now found a process
which is significantly cheaper due to avoiding the loss of HCl and due
to energy savings in the HCl removal and recovery. In this process the
biomass is contacted with aqueous hydrochloric acid for a short time to
dissolve the biomass, while this process does not involve complete
hydrolysis of the biomass. According to this process, oligosaccharides
containing products can be obtained which appear to be well fermentable
into, and useful for the fermentative production of a variety of useful
products. For example these oligosaccharides appear to be convertible
into glucose by cellulase enzymes at a surprisingly high rate as com
pared with the direct conversion of the milled biomass. Fermentation of
the oligosaccharides into ethanol at high conversion rates appears to
be possible.

Further the use of the oligosaccharide fraction directly for the micro bial production of e.g. penicillin has proved to be feasible. In addi tion enzymes (e.g. cellulase and amyloglucosidase) can be produced. Moreover the ologisaccharides may be used as substitution of or addi tion to cattle feed, e.g. non ruminant cattle such as pigs, in order to save costs. More advantageously the lignin containing products may be applied to provide structure to the feed.

As an outline the process can be described as follows:

a. the biomass is contacted with a highly concentrated aqueous (or gaseous) hydrochloric acid solution until substantially all of the cellulose and hemicellulose have been dissolved, 5 45 minutes at 30°C; lignin and other materials insoluble in highly concentrated HCl can optionally be removed by centrifugation or filtration in this stage

b. the reduction of the hydrochloric acid concentration under reduced pressure with optionally concurrent precipitation of the oligosac charides containing product until the concentration of the water hy drochloric acid azeotrope is reached and recycling of the gaseous hydrochloric acid and condensed azeotrope of hydrochloric acid to stage a.

c. separation of the precipitated oligosaccharides containing product from the aqueous hydrochloric acid solution and recycling of the latter for use in stage a.

The savings on energy are significant compared with the process as des cribed in a European patent application (Battelle) due to the reduction of the hydrochloric acid concentration for instance to 25% by weight as calculated in Table I.

Table I

Biomass to liquid ratio	Energy required for removal of HCl per kg sugars (MJ/kg)		
	Fuming HCl, total hydrolysis (reduction to 1% HCl)	This process (reduction to 25% HCl)	European patent application 52896
1:4	12.60	1.52	11.6
1:3	9.48	1.14	8.7
1:2	6.32	0.78	5.8
1:1	3.16	0.38	2.9

Since the hydrogen chloride recovered in the evaporation stage and the supernatant, containing azeotropic hydrochloric acid, are used to ini tiate the next digestion cycle, the only acid lost would be that adhe ring to the oligosaccharides containing precipitate and lignin but this can be recovered satisfactorily by a judicious choice of a washing method. The loss of HCl will be 0.14 tonne HCl/tonne ethanol formed in the fermentation process which is significantly low compared with the process as described in European patent application 52896 in which the loss of HCl is 0.30 1.0 tonne HCl/tonne ethanol. Probably even more HCl is lost in the latter process due to tightly bound HCl to lignin, which is not accounted for.

The next detailed experiment shows that cellulose is easily converted to glucose by Trichoderma viride cellulase. In the laboratory the lig nin is difficult to separate from the dissolved cellulose. Therefore we choose to leave the lignin in the precipitated cellulose fraction.

The dissolved hemicellulose is found in the HCl supernatant. The concentration of the sugars (hexoses and pentoses) will build up to about 40% (in the recycling stream). These monosugars will be recovered from a bleed of this recycling stream.

The results indicated in Table II show that initial cellulase hydrolysis rates can be obtained that are 26 to 34 times faster than for a sample of milled poplar wood. To obtain this result concentrated hydrochloric acid (40 ml) was saturated with hydrogen chloride gas at 20°C and then 4.0 g of knife milled poplar wood was added with overhead stirring. Hydrogen chloride gas was passed in for 20 minutes at 20°C with stirring. The green black slurry was then concentrated under reduced pressure with a bath temperature of 25°C until the concentration of hydrochloride acid was reduced to 22 23% by weight. Centrifugation gave a pellet of the oligosaccharides containing products and lignin, which was washed with 2 x 9 ml of water. The pellet was neutralised with sodium bicarbonate, washed with water and freeze dried.

The supernatant and primary washings were made up to 40 ml with concentrated hydrochloric acid and saturated with hydrogen chloride gas at 20°C with stirring to start a second cycle with 4.0 g of poplar wood digestion as for the first cycle. In this way 8 cycles of poplar wood digestion were carried out, as shown in Table II.

The product of this process can be used in many applications such as enzyme as well as antibiotic fermentations, and in various feed mixtures. Some examples will be given in the presentation.

Table II

Cycle	Weight of precipitated* oligosaccharides containing products + insoluble products (lignin) (g)	% Yield on dry starting wood	Relative initial rate in cellulase assay
1	2.24	59	25.6
2	2.72	71	33.9
3	2.13	56	32.6
4	2.32	61	34.1
5	2.07	54	32.9
6	2.23	58	30.2
7	2.15	56	29.8
8	2.85	75	31.2
milled poplar wood			1.0

* corrected for handling losses

THE COMMON AGRICULTURE POLICY AND NEW USES
FOR LIGNOCELLULOSIC MATERIAL

W. FLOYD

Directorate-General for Agriculture

Summary

During 1985 the Commission undertook a wide consultation on the
future directions that the CAP ought to take to resolve the evident
problems facing it. The policy guidelines which emerged are relevant to
the development of new uses for lignocellulosics in a number of ways.

Most importantly, one of the six main objectives is precisely the
development of industries which process agricultural produce. This paper
briefly outlines these objectives and the methods that you can expect the
Commission to use to reach them.

Objectives

The Commission's guidelines were published in December last year. The
objectives had to correspond with those of the Treaty of Rome, and as far
as possible with the wishes expressed during the consultation procedure.
Those that emerged were :

- gradually to reduce production in the sectors which are in surplus and
  to alleviate the resulting burden on the taxpayer;

- to increase the diversity and improve the quality of production by
  reference to the internal and external markets and the desires of
  consumers;

- to deal more effectively and systematically with the income problems of
  small family farms;

- to support agriculture in areas where it is essential for land use
  planning, maintenance of the social balance and protection of the
  environment and the landscape;

- to make farmers more aware of environmental issues;

- to contribute to the development in the Community of industries which
  process agricultural produce, and thus involve agriculture in the
  profound technological changes which are taking place.

These priorities are not just a passing response to the present problems of surplus production and budgetary cost. They are guidelines which will be followed for as far into the future as one can see. The interesting part is to see just how.

The Methods

The methods can be conveniently divided up between 4 main policy areas. These are prices and markets, structural policy, environmental policy, and external relations.

In the area of prices and markets, a purely price policy would have been too drastic. Quotas were not acceptable either. So the Commission chose a package of measures, including price restraint, emphasis on qulity, and co-responsibility. As is already the case with sugar, this can be used to cover the costs of finding an outlet for the commodity in question. The farming organisations want to have a say in how the funds raised by a co-responsibility levy would be spent. One can expect that there would be a lot of support for using them to promote new uses of their products. Let it be said, though, that any further expansion of production must henceforth be justified by a real — and unsubsidised — expansion in demand.

The Commission will be very vigilant as regards any national aids which can, directly or indirectly, cause an increase in production. A proliferation of these would be tantamount to a "renationalisation" of the agricultural policy, and will not be tolerated. The same goes for individual national legislation concerning the fitness for use of agricultural produce.

Forestry is recognised as having a specific part to play in the changes expected for land use patterns. Although some Member States still resist the adoption of a Community Forestry Policy as such, the Commission is proposing an enormously increased programme of activity in this sector.

Although much is heard about the proposed new regimes for sugar and starch — proposals designed to encourage their industrial use within the Community at a competitive price — the principle could have wider application. It only requires the processing industry to make a convincing case with a sound economic future.

There will be new initiatives in the Community's structural policies for agriculture. These will concentrate on improving conditions by cutting costs rather than by promoting greater production. In fact, the scope of measures to help the less-favoured regions, mountain, and hill-farming, will exclude any intensification or net increase in production. The emphasis will be on rationalisation, improved organisation, and marketing, and qualitative improvement of the output.

Environmental policy will be increasingly integrated into agricultural policy - as it is in all other areas of Community policy as well. The effect of this will be that assistance will go hand in hand with a respect for the environment, and of course the health of the consumer. In this way, concern for the environment will have a more preventive, and thus more effective nature.

Finally, external relations will be becoming more important. In the coming round of GATT negotiations much attention will be devoted to agricultural products. To be in a good position for this round, we simply must set our own house in order first. This is not just a matter of removing our existing stocks, but of curtailing the excess production. If we can do this, and only if we can do this, will we be in a strong position vis-a-vis trading partners who may have surplus stock disposal problems, and which can interfere with our markets.

Conclusion

There are three messages in this for those who are looking for new uses of lignocellulosic material, and they are encouraging. Firstly, the Commission is absolutely in favour of these new developments so long as they respect the environment, and avoid direct or indirect surplus production. Secondly, the proponents should not have to lay their plans on the assumption that non-viable production will continue to be supported within the Community. And finally, as Community agriculture becomes more adapted to market realities, the farmers will be actively seeking ways to reduce their unit costs of production.

DISCUSSION

Session I

Degradation in animals, related to structure of the tissue

Degradation in industrial processes, related to structure of the tissue

After removal of only a part of the lignin the digestibility of the cell-walls already increases considerably. The effect of such a delignification on the content of reducing sugars is limited. The significance of lignin content as an indicator of low digestibility is overestimated; there are different relations between the lignin content and digestibility in different plants. When products are treated; the properties of what is called lignin change on treatment. The structure of cellulose itself is more important as to digestibility than the lignin content. For instance, in a collection of 88 straw samples a linear regression between digestibility and swelling has been found ($r = 0,73$). It is therefore to be expected that the ease with which enzymes and bacteria can penetrate in the cell wall is the main factor determining the extent of digestibility. Availability for bacterial degradation is influenced by the thickness of the cell walls, extent of lignification and encrustment by "warthy layers". During delignification there is a decrease in the content of hemicelluloses. On steam treatment there is an increase in the measured lignin content due to condensation, still there is an increase in digestibility at the same time. Plant cell types differ because of light, soil and climate conditions during their growth period and in nodes, internodes and leaves they are not equal. Soil strongly determines the silica content but hardly influenced digestibility. The silicium effect on digestibility in the tropics is mainly caused by the dilution effect on dry matter, but soil pH may influence the uptake of silicum, as is known that in the tropics acid soils cause high Si-intake. Si-intake differs between sandy, light and clay soils.

In plants grown under high temperature conditions the lignin content was found to be twice as high as at low temperature conditions and digestibility showed the opposite effect. But the digested cellwalls were not the same; on maturation of the plants, thickening of the warthy layer is accompanied by decreasing digestibility of certain plant cells. The development of this warthy layer depends on the type of plant, plant parts and growing conditions. According to van Soest cell-contents are digestible for about 98% but whole cells have been observed in faeces. It was also observed that some of the N in the faeces was non-digested protein inside intact cells. Probably a major condition for digestion is that cells are not encrusted by warthy layers.

On heating the production of Maillard reaction products is advanced as well as condensation of the lignin and polymerisation of the carbohydrates by dehydration.

First the free water is removed then irreversable dehydration of the cristal water occurs. When less than 8% moisture is left in plantmaterials these have irreversably changed. On freeze drying the original structure will be maintained.

On grinding the animals' voluntary intake and the rate of passage of the fed through the GIT increase but the apparent digestibility of the slower digestible cell walls will decrease.

Grinding procedures influence the cristallinity as has been shown with X-ray using different mills and so the possible penetrations of enzymes. Penetration can be improved by swelling by treatment with acids. Heat and steam increase the cristallinity of the residue, but this effect is only at the surface which is profitable to other industrial applications.

Swelling and degradation by sodium hydroxide or caustic soda is usually too expensive for feed applications and has also negative health and environment aspects. The application of ammonia and urea or other ammonia releasing salts has been increased but is also insufficiently applicable under local tropical conditions because of transport and water needs, moreover it could also be used as a fertilizer.

Oxidising agents like sulphurdioxide (SO_2), peroxides (H_2O_2, alkaline) and ozon (O_3) are not applicable on farm sites. In situ H_2O_2 production already economically used for bleaching of pulp may be available for farmers also. Sodium hydroxide and peroxide treatment of straw produce a high quality pulp for paper production, which however is too expensive for animal feed.

Enzymes have a limited applicability as yet. Lignases are redox enzymes and not easily applied as a free catalyst, although combined with peroxides positive expectations for future application have been expressed.

For grass silage cellulase and gluco-oxidases are applied in combination, not so much for improvement of the digestibility, but for rapid stabilisation, i.e. a low pH.

For the moment enzymes are still expensive and in low concentrations only applicable at increased environmental temperature and reaction time. Fungi have good expectations for the moment although controlled conditions and profitable applications for by-products are needed.

Future attention is wanted for combinations of treatments for large and small scale applications. More clearness must be obtained on the economic feasability of processes to be applied on the feedstocks available.

Session II

Feeding value as related to the composition of the material fed.

Merits of the various industrial processes– a comparison for raw materials.

Increase of animal production is of great importance to developing countries. This has not been achieved because of the availability of low quality materials as the main source of feed.

Because of the fact that the volume of the forestomachs is related to the live weight of the animals (LW^1) whereas the animals' maintenance requirement is related to $LW^{\frac{3}{4}}$ it makes sense when only low quality feeds are available to use large size animals. Still, feeds of very low quality as such can only be used for ruminants requiring little, i.e. non-producing animals.

Digestibility of feeds explains about 70 percent of their net energy content while their ingestibility, the amount eaten voluntarily, explains the major part of the remaining 30%. This latter part is mainly related to the rate of degradation in the forestomachs by rumen microbes and rumination. Both digestibility and degradation rate can be measured by nylon bag and in vitro techniques. Of course these methods predict for standardized circumstances only, often the maintenance feeding level and secondary relationships must be derived to predict for instance at other production levels.

Feed intake is influenced by the physical conditions of the fibrous feeds, as discussed already in session I.

Without losing the long particle size for ruminants the access for bacteria and enzymes to speed up degradation rate might be improved by treatments like steam explosion treatment. Particle size affects passage rate, but also small particles are retained to some extent from transport to the true stomach by adhering gas. Degradation rate of well digestible material is about 8 percent per hour while the passage rate is about half the digestion rate – 4,5 percent per hour. The effect of treatment on net energy content is sometimes even negative. Many treatments cause lignin condensation decreasing the digestibility. This is also the reason that often the by-products of production of pulp for paper cannot be used for livestock feed. Fungal treatment of straw in solid state fermentors under controlled circumstances as to water, oxygen, carbondioxide and temperature using well-selected fungi varieties is possible with not more than 10 to 15 percent loss of organic matter, increasing the digestibility somewhat.

The procedure still faces up scaling problems, is expensive and is never applicable at farm scale.

Pulp of olive and grapes-mare contain very high lignin contents (up to 50%) and insufficient energy in carbohydrates to allow for profitable treatments. Even after alkaline treatment these materials are low in digestible organic matter and available nitrogen.

Alternative industrial applications making use of their composition should be studied.

Session III

Feed resources, available in Europe and criteria for utilization by farm animals.

Actual and future industrial applications of renewable resources.

The shortage in pulp for paper from european sources is sufficiently supplemented by imports from America. Therefore expansion and intensification of the european forestry is very dependent on the market and politics.

In Europe about 14 million tons of straw per year are available. In Denmark up to 30% of straw pulp is included in quality paper pulp. It is the silica content of straw which causes fairly high loss of acid catalyst; also softwood needs more catalyst than hardwood.

Only 45% of the straw is retained as cellulose, the remainder is yet abandoned. Because the leaves contain a cellulose with a low quality for paper pulp these have to be separated from the stem and are available as a feed source or for the production of chemicals.

Softwood has lower strength properties for paper pulp and delignification is more difficult because its lignin is more reactive. The acid lowers the degree of polymerisation, the fibres are cracked so that a finishing treatment is needed. One ton of straw produces about 80 kg of lignin. This material is only used as fuel since at present the costs of separating this lignin are too high.

Organosolv pulping which can be done in small plants, produces fairly pure by-products which can be used for viscose rayon, ethers etc.

In Europe, dextrose production from cellulose cannot compete with starch, therefore waste materials from this process are either burned or used as a cattlefeed. Also the GB-Shell process produces fairly pure compounds used for ethanol production through enzymatic degradation, by microorganisms. The cellulose fraction has a low polymerisation degree (10-100). Only for quickly renewable resources, i.e. for instance rapidly growing eucalyptus trees, the process is economical.

Straw and its derived compounds have only limited applications yet:

- livestock   = feedstuff, bedding

- fibre   = fibre particle board, energy

- cellulose   = paper, cellulose derivatives, sugars, ethanol

- hemicellulose   = furfural, furan derivatives, xylose, xylitol, single

  cell protein, amino acids

- lignin   = coloured polymethane foam, phenols, resins, pharma-

  ceuticals, controled release matrix, absorbens, anti-

  oxidants, bioregulation, diesel fuel, fuel

- waxes   = potential: 600 tons/yr in Europe (1 million ECU/yr)

There is a great need for knowledge on derivatives of C5-sugars making use of the special properties of the C5-structure. Their application depends on the quality of the by-products. Therefore quality demands for applications of the by-products from the pulp production must be defined so that the pulping processes can be adapted.

The present economic situation is a bad guide in selection topics for research and development. But from biotechnology and other recent developments new products and applications may be expected.

For production of cellulose fibre also plants like flax, hemp, eucalyptus, kenaf, agave, elephant grass and bamboo (donax reed) are available. Plant breeders and biotechnologists should work on a plant for Europe (for instance in Eureka) which is high in cellulose and low in hemicellulose and lignin.

The present situation in crop production will only change depending on the

local market prices.

It is doubted if the growth of biomass for the production of organic chemicals only is economically feasable. Especially the market for ethanol is flooded and derivation processes are needed.

RECOMMENDATIONS

* A good coördination of activities regarding lignocellulosics of COST, EC and OECD is recommended.

* Multi-purpose rather than single-purpose use of lignocellulosics should be considered.

* Lignocellulosics are heterogeneous as to structure and composition, not only due to plant species and variety and due to plant part but also within tissue sections.
  This has important implications for their use.

* Attention should be paid to new promising analytical techniques which give more insight in rate and extent of degradation of lignocellulosics in the ruminant (modern types of chemical/physical analysis; incubation of tissue sections with rumen fluid of various length before and after a treatment in combination with staining and light- and electron microscopy; in vitro incubations of ground lignocellulosics of various duration with rumen fluid etc.)

* Mode of actions of various technological treatments for improving lignocellulosics as a feed for ruminants are still not wholly understood. Not all are effective, some have hazards for human health and/or the environment.

* For optimal use of lignocellulosics as a feed for ruminants three aspects

are important:

1) rate of degradation by rumen microbes - in view of voluntary intake - and digestibility - in view of energy and protein in feed value -,

2) optimal conditions for microbes in the forestomachs - in view of high ration intake -

These optimal conditions are a.o.: no shortage of N, S, P, energy, pH>6, some but not too much easily fermentable material, all this for 24h of the day,

3) The ruminant itself should not have shortages as to one or more essential nutrients at the tissue level; if so, maximum rate of production cannot be achieved and as a consequence appetite will be depressed.

* Degradation of lignocellulosics with special fungi that decompose lignin leads only under certain conditions to sufficient increase in digestibility. By present knowledge this process seems only economically feasible if combined with production of edible mushrooms.

* Silica in straw and other plants makes the lignocellulosics less suited for use for paper production; more information on factors determining uptake of silica by the plant (variety/species; soil type; climate/rate of evaporation) is needed.

* There is a shortage of long-fibre plants for pulp production in Europe. New pulping procedures are needed to enable the use of plants other than wood and straw, like: eucaliptus, flax, kenaf, hemp, elephant grass and bamboo (donax reed).

* Plant breeders should work on low cost plants with long fibres, high cellulose, low lignin. For this reason it must be defined which is the ideal plant for the pulping process or for animal feed.

* The European Association of Animal Production (EAAP) collects information on lignocellulosics available for animal feeds. Also for industrial application such information is needed.

* Processes which produce relatively clean hemicelluloses and lignins or their mono-/oligomers should be studied and developed.

* Pre-treatments, before the pulping process, with fungi or other chemicals to release lignin are needed for the production of well-defined oligomers, suited for new applications of the polyphenols.

* Hemicelluloses will become available. Applications making use of compounds posessing the C_5 structure are needed; such chemicals are especially needed in bulk.

L I S T   O F   P A R T I C I P A N T S

AGUILERA, J.F.
Estación Experimental del Zaidin
Dept Animal Physiology
Prof. Alberada 1
E - 18008 GRANADA

BESLE, J.M.
Laboratoire des Aliments
INRA-CRZV de Theix
F - 63122 CEYRAT

BROGAN, J.C.
An Foras Taluntais
Dunsinea Research Centre
Castleknock
IRL - DUBLIN 14

CHENOST, M.
Laboratoire des Aliments
INRA-CRZV de Theix
F - 63122 CEYRAT

CHESSON, A.
Rowett Research Institute
Bucksburn
UK - ABERDEEN AB2 9SB

DEGER, H.-M.
Zentralforschung II/Biotechnik
i. Hs. Hoechst AG
Postfach 80 03 20
D - 6230 FRANKFURT/MAIN 80

DELORT-LAVAL, J.
Directeur de Recherche à l'INRA
Chemin de la Géraudière
F - 44072 NANTES Cedex

EDEL, E.
M.D. Organocell
Planeggerstraße, 38
D - 8000 MÜNCHEN 60

ENGELS, F.M.
Landbouwhogeschool
Plantcytology and Morphology
Arboretumlaan, 4
NL - 6703 BD WAGENINGEN

FLOYD, W.
Directorate-General Agriculture
Commission of the European
Communities
200, rue de la Loi
B - 1049 BRUXELLES

HÄRNMERLI, ST.
Swiss Federal Institute of
Technology
RTH-Hönggerberg
CH - 8093 ZÜRICH

HARTLEY, R.D.
The Grassland Research Institute
Hurley
UK - MAIDENHEAD, Berks SL6 5LR

KLEINHANS, W.
Directorate-General Science,
Research and Development
Commission of the European
Communities
200, rue de la Loi
B - 1049 BRUXELLES

KREMER, P.
Dechema
Postfach 97 01 46
D - 6000 FRANKFURT 97

SACHETTO, J.-P.
Batelle
7, rue de Drize
CH - 1227 CAROUGE (Genève)

LEBZIEN, P.
Institut für Tierernährung (FAL)
Bundesallee, 50
D - 3300 BRAUNSCHWEIG

SANDERS, J.
Gist Brocades NV
Postbus 1
NL - 2600 MA DELFT

PANCIROLI, A.
Zootechnical Institute
Via S. Giacomo, 11
I - 40126 BOLOGNA

VAN ES, A.J.H.
Institute for Livestock Feeding
and Nutrition Research (IVVO)
P.O. Box 160
NL - 8200 AD LELYSTAD

PETERSEN, U.
Bundesministerium für Ernährung,
Landwirtschaft und Forsten
Postfach 14 02 70
Rochusstraße, 1
D - 5300 BONN 1

VAN DE VOORDE, L.
Lab. voor algemene en toegepaste
microbieële ecologie
Fac. Landbouwwetenschappen RUG
Coupure Links, 653
B - 9000 GENT

POHL, R.
Maizena Krefeld
Postfach 90 80
D - 4150 KREFELD/LINN

VAN DER MEER
Institute for Livestock Feeding and
Nutrition Research (IVVO)
P.O. Box 160
NL - 8200 AD LELYSTAD

PULS, J.
Bundesforschungsanstalt für Forst-
und Holzwirtschaft
Postfach 80 02 10
D - 2050 HAMBURG 80

ZADRAZIL, F.
FAL
Bundesallee, 50
D - 3300 BRAUNSCHWEIG

REXEN, F.
Carlsberg Research Lab.
Gamle Carlsbergvej. 10
DK - 2500 COPENHAGEN

RIJKENS, B.A.
Institute for Storage and Processing
of Agr. Prof. (IBVL)
P.O. Box 18
NL - 6700 AA WAGENINGEN

RIZZI, L.
Zootechnical Institute
Via S. Giacomo, 11
I - 40126 BOLOGNA